SOLUTIONS MANUAL

for

Ebbing/Wentworth
INTRODUCTORY CHEMISTRY

prepared by

David Bookin
MOUNT SAN JACINTO COLLEGE

HOUGHTON MIFFLIN COMPANY   BOSTON   TORONTO

GENEVA, ILLINOIS   PALO ALTO   PRINCETON, NEW JERSEY

Copyright © 1995 by Houghton Mifflin Company. All rights reserved.

No part of this work may be reproduced or transmitted in any form or by any means, electronic or mechanical, including photocopying and recording, or by any information storage or retrieval system without the prior written permission of Houghton Mifflin Company unless such copying is expressly permitted by federal copyright law. Address inquiries to College Permissions, Houghton Mifflin Company, 222 Berkeley Street, Boston, MA 02116.

Printed in the U. S. A.

ISBN: 0-395-46629-6

123456789-B-99 98 97 96 95

CONTENTS

Preface / v

1. **INTRODUCTION TO CHEMISTRY** — 3
   Answers to Questions to Test Your Reading 3

2. **MEASUREMENT IN CHEMISTRY** — 5
   Solutions to Exercises 5 / Answers to Questions to Test Your Reading 7 /
   Solutions to Practice Problems 9 / Solutions to Additional Problems 14

3. **MATTER AND ENERGY** — 17
   Solutions to Exercises 17 / Answers to Questions to Test Your Reading 18 /
   Solutions to Practice Problems 20 / Solutions to Additional Problems 22

4. **ATOMS, MOLECULES, AND IONS** — 24
   Solutions to Exercises 24 / Answers to Questions to Test Your Reading 25 /
   Solutions to Practice Problems 27 / Solutions to Additional Problems 30

5. **CHEMICAL FORMULAS AND NAMES** — 32
   Solutions to Exercises 32 / Answers to Questions to Test Your Reading 33 /
   Solutions to Practice Problems 34 / Solutions to Additional Problems 36

6. **CHEMICAL REACTIONS AND EQUATIONS** — 38
   Solutions to Exercises 38 / Answers to Questions to Test Your Reading 43 /
   Solutions to Practice Problems 45 / Solutions to Additional Problems 52

7. **CHEMICAL COMPOSITION** — 56
   Solutions to Exercises 56 / Answers to Questions to Test Your Reading 61 /
   Solutions to Practice Problems 62 / Solutions to Additional Problems 88

8. **QUANTITIES IN CHEMICAL REACTIONS** — 99
   Solutions to Exercises 99 / Answers to Questions to Test Your Reading 103 /
   Solutions to Practice Problems 105 / Solutions to Additional Problems 117

9. **ELECTRON STRUCTURE OF ATOMS** — 125
   Solutions to Exercises 125 / Answers to Questions to Test Your Reading 126 /
   Solutions to Practice Problems 128 / Solutions to Additional Problems 130

iv ■ CONTENTS

10. PERIODIC PROPERTIES OF THE ELEMENTS — 132
Solutions to Exercises 132 / Answers to Questions to Test Your Reading 132 /
Solutions to Practice Problems 134 / Solutions to Additional Problems 135

11. CHEMICAL BONDING — 137
Solutions to Exercises 137 / Answers to Questions to Test Your Reading 142 /
Solutions to Practice Problems 145 / Solutions to Additional Problems 149

12. THE GASEOUS STATE — 153
Solutions to Exercises 153 / Answers to Questions to Test Your Reading 161 /
Solutions to Practice Problems 163 / Solutions to Additional Problems 173

13. LIQUIDS, SOLIDS, AND ATTRACTIONS BETWEEN MOLECULES — 176
Solutions to Exercises 176 / Answers to Questions to Test Your Reading 178 /
Solutions to Practice Problems 181 / Solutions to Additional Problems 188

14. SOLUTIONS — 190
Solutions to Exercises 190 / Answers to Questions to Test Your Reading 194 /
Solutions to Practice Problems 200 / Solutions to Additional Problems 212

15. REACTION RATES AND CHEMICAL EQUILIBRIUM — 219
Solutions to Exercises 219 / Answers to Questions to Test Your Reading 220 /
Solutions to Practice Problems 223 / Solutions to Additional Problems 228

16. ACIDS AND BASES — 230
Solutions to Exercises 230 / Answers to Questions to Test Your Reading 232 /
Solutions to Practice Problems 236 / Solutions to Additional Problems 243

17. OXIDATION-REDUCTION REACTIONS — 245
Solutions to Exercises 245 / Answers to Questions to Test Your Reading 251 /
Solutions to Practice Problems 254 / Solutions to Additional Problems 266

18. NUCLEAR CHEMISTRY — 274
Solutions to Exercises 274 / Answers to Questions to Test Your Reading 275 /
Solutions to Practice Problems 278 / Solutions to Additional Problems 283

19. ORGANIC CHEMISTRY — 285
Solutions to Exercises 285 / Answers to Questions to Test Your Reading 289 /
Solutions to Practice Problems 294 / Solutions to Additional Problems 304

20. BIOCHEMISTRY — 310
Solutions to Exercises 310 / Answers to Questions to Test Your Reading 311 /
Solutions to Practice Problems 315 / Solutions to Additional Problems 320

APPENDIX A — 322
Solutions to Exercises 322

APPENDIX B — 323
Solutions to Exercises 323

PREFACE

This complete Solutions Manual provides worked-out answers to all of the problems that appear in *INTRODUCTORY CHEMISTRY* by Ebbing and Wentworth. This includes detailed, step-by-step solutions for all in-chapter Exercises as well as for the Practice Problems and Additional Problems that appear at the end of the chapters. Also provided are answers to all of the Questions to Test Your Reading.

Please note the following:

Significant figures: The answer is first shown with 1 to 2 nonsignificant figures and no units, and the least significant digit is underlined. The answer is then rounded off to the correct number of significant figures with units.

Great effort and care have gone into the preparation of this manual. The solutions have been checked and rechecked for accuracy and completeness several times. I would like to thank John Goodinow who helped check the solutions for accuracy.

D. B.

SOLUTIONS MANUAL

for

Ebbing/Wentworth
INTRODUCTORY CHEMISTRY

1. INTRODUCTION TO CHEMISTRY

■ Answers to Questions to Test Your Reading

1.1 Three examples, with properties given after each, are: 1) copper: firm solid with reddish color; 2) water: colorless and liquid, and 3) oxygen: colorless and gaseous.

1.2 Three examples, with their descriptions given after each are: 1) wood burning in air: the flame is bright and produces smoke, gases, and ash; 2) baking soda and vinegar reacting: they form a clear solution and bubbles of carbon dioxide gas, and 3) copper and nitric acid reacting: they form a blue-green solution and a red-brown gas.

1.3 Two reactions and their energy changes are: 1) wood burning in oxygen to evolve heat energy and light energy, and 2) baking soda and vinegar reaction to evolve some heat energy and sound energy (fizzing of gas).

1.4 Chemistry is the science concerned with describing and explaining the different forms of matter and the chemical reactions, with energy changes, of the various forms of matter.

1.5 Chemistry is important to these other sciences:
Biological study and research both involve chemical processes that occur in a cell. In physics, the development of superconductors depends on the chemical synthesis of new materials such as ceramic materials that will superconduct closer to room temperature. In geology, geologists called geochemists study chemical reactions that occur within or on the earth's surface.

1.6 Some applications are: the chemical manufacture of colored pigments and dyes, obtaining perfumes from natural materials by extraction, and the manufacture of pottery and metals.

1.7 Bronze was discovered when copper ore and tin ore were heated with charcoal. Glass was first produced by heating a mixture of sand, limestone, and plant ashes.

1.8 Aristotle viewed matter as having four properties or qualities: hot, cold, dry, and wet. These were represented as the corners of a square: prime matter containing two properties on adjacent corners of this square would be one of the four possible elements of fire, air, water, and earth.

1.9 Democritus believe matter consisted of very small particles or atoms which were too small to be seen and which could not be cut into smaller parts.

1.10 Priestley discovered that glowing wood would burst into brilliant flame when put into a bottle of oxygen, and that a mouse could live much longer in the same volume of oxygen as air.

1.11 Lavoisier's experiments on combustion, etc. were characterized by measurements of the amounts of reactants and products involved. He weighed these in terms of grains and measured volume in terms of cubic inches.

1.12 Lavoisier explained the combustion of mercury as a reaction of mercury with oxygen gas to form the compound mercury(II) oxide.

1.13 The atomic theory of chemistry is associated with Dalton. This theory is based on the existence of matter in the form of atoms, and the idea that atoms of a given kind have a definite mass, as a measure of the quantity of matter. Dalton was able to explain the quantitative results that chemists obtained in their experiments.

1.14 a) An experiment is the observation by a scientist of some natural phenomenon occurring under controlled conditions.

b) A law is a simple generalization obtained from an experiment.

c) A hypothesis is a tentative explanation of a law.

d) A theory is a tested explanation of some body of natural phenomena.

1.15 The scientific method advances a field by means of the testing of various hypotheses. The results of the experiment either confirm a particular hypothesis or lead to its modification or amplification. More experiments are usually done, leading first to a better hypothesis, and ultimately to a theory.

2. MEASUREMENT IN CHEMISTRY

■ Solutions to Exercises

Note on significant figures: Starting with Exercise 2.4, when the final answer is written for the first time, it is written with one nonsignificant figure. The least significant digit is also underlined. The final answer is then rounded to the correct number of significant figures but the least significant digit is no longer underlined.

2.1 In parts a, b, and e, note that the beginning zeroes are not significant, and do not appear in the answer. In parts b and c, the terminal zeroes are significant, and do appear in the answer. The answers to each part are:

    a. 6.9252×10^{-3} b. 6.300×10^{-2} c. 8.0200×10^{2}
    d. 9.002×10^{3} e. 7.07×10^{-1}

2.2 Move the decimal point the number of places indicated by the exponent. Write zeroes in front of numbers that have negative exponents, as for part a.

    a. 0.0004834   b. 6250.   c. 4.89

2.3 Remember that the beginning zeroes are not significant, and terminal zeroes are significant only where there is a decimal point.

    a. 4 sig. figs.   b. 4 sig. figs.   c. 4 sig. figs.
    d. at least 2 sig. figs.   e. 5 sig. figs.

2.4 Starting with this exercise, the first time the final answer is written, it is written first with one nonsignificant figure. The least significant digit is also underlined. The final answer is then rounded to the correct number of significant figures.

    a. 97.0̲8 = 9.71×10^{1} b. 28.2̲6 = 2.83×10^{-1}
    c. $3.08̲2 \times 10^{1}$ = 3.08×10^{1} d. $3.2̲05 \times 10^{0}$ = 3.2
    e. $3.62̲3 \times 10^{-1}$ = 3.62×10^{-1}

CHAPTER 2

2.5 Do the calculation inside the parentheses first, leaving the intermediate answer in the calculator for the next step. Review the directions at the start of this section for the way significant figures are reported below.

  a. $29\underline{9}.97 = 3.00 \times 10^2$ [Note that $102.5 \times (5.62 - 3.34) = 23\underline{3}.7$ by the mul/div rule.]

  b. $7.\underline{3}1 = 7.3$ [Note that $8.64/3.2 = 2.\underline{7}0$ by the mul/div rule.

2.6 a. $2.58 \text{ g} \times 10^{-3} \text{ kg/g} = 2.58 \times 10^{-3} \text{ kg}$

  b. $55.4 \text{ cm} \times 10^{-2} \text{ m/cm} = 5.54 \times 10^{-1} \text{ m}$

  c. $7.11 \text{ L} \times 10^3 \text{ mL/L} = 7.11 \times 10^3 \text{ mL}$

2.7 a. $2.45 \text{ ns} \times 10^{-9} \text{ s/ns} \times 10^3 \text{ ms/s} = 2.45 \times 10^{-6} \text{ ms}$

  b. $7.38 \text{ kg} \times 10^3 \text{ g/kg} \times 10^3 \text{ mg/g} = 7.38 \times 10^6 \text{ mg}$

  c. $5.02 \text{ } \mu\text{L} \times 10^{-6} \text{ L/}\mu\text{L} \times 10^3 \text{ mL/L} = 5.02 \times 10^{-3} \text{ mL}$

  d. $89.7 \text{ cm} \times 10^{-2} \text{ m/cm} \times 10^{-3} \text{ km/m} = 8.97 \times 10^{-4} \text{ km}$

  e. $6.11 \text{ ms} \times 10^{-3} \text{ s/ms} \times 10^6 \text{ } \mu\text{s/s} = 6.11 \times 10^3 \text{ } \mu\text{s}$

2.8 $6456 \text{ lb} \times 1 \text{ ton}/2000 \text{ lb} = 3.228 \text{ ton}$

2.9 In the temperature conversion from absolute to Celsius, use this formula:

$t_{°C} = t_K - 273$

$t_{°C} = 296 - 273 = 2\underline{3}.0 = 23° \text{ C}$

2.10 a. In the temperature conversion from Celsius to Fahrenheit, use this formula:

$t_{°F} = 1.8 \, t_{°C} + 32$

$(1.8 \times 35) + 32 = 9\underline{5}.0 = 95° \text{F}$ (note that the ".0" is not significant)

  b. In the temperature conversion from Fahrenheit to Celsius, use this formula:

$t_{°C} = (t_{°F} - 32) \div 1.8$

$(102 - 32) \div 1.8 = 3\underline{8}.88 = 39° \text{ C}$

MEASUREMENT IN CHEMISTRY ■ 7

2.11 The density is calculated as follows:

density = mass ÷ volume = (38.53-24.59 g) ÷ 4.50 cm^3 = 3.0$\underline{9}$7 = 3.10 g/mL

2.12 Density = mass ÷ volume; rearranging gives volume = mass ÷ density, and

$$3.14 \times 10^4 \text{ g} \times \frac{1 \text{ cm}^3}{21.45 \text{ g}} = 1.4\underline{6}3 \times 10^3 = 1.46 \times 10^3 \text{ cm}^3$$

2.13 Density = mass ÷ volume; rearranging gives volume = mass ÷ density, and

$$\frac{1.50 \text{ g}}{1 \text{ mL}} \times 8.4 \text{ mL} = 1\underline{2}.6 = 13 \text{ g}$$

■ Answers to Questions to Test Your Reading

2.1 In measuring a the length of a rod, the measurement process will involve comparing that length with a standard measurement device, such a yardstick or ruler. You place the rod with one end at the zero inches, and read the length at the point on the yardstick or ruler where the other end is located. All such measurements involve a comparison with an established standard like the yardstick for one yard or the ruler for one foot. The units are important to be sure that whoever reads the written result knows what the numbers mean.

2.2 When the student pressed the equal sign, the calculator automatically converted the readout to scientific notations. A calculator will not do this all of the time, but this particular calculator was programmed to read out in the scientific notation <u>mode</u>.

2.3 The answer is 10^{-4} or 0.0001.

2.4 The four base units are meter (length), kilogram (mass), second (time), and kelvin (temperature).

2.5 a. kilo   b. centi   c. micro   d. mega

2.6 Yes, there is a difference. Heat is a form of energy, but temperature is a quantity you ascribe to a given point in a sample, and does not have to do with the total sample.

2.7 The freezing points are 0°C, 32°F, and 273 K.

2.8 This means that it is a scale which has zero for its lowest temperature.

2.9 There is one cubic decimeter and 1000 cubic centimeters in one liter.

8 ■ CHAPTER 2

2.10 The uncertainty is ±1 in the last digit in any measurement. Since one mark on a meter stick is 0.001 m, or 1 mm, the uncertainty of either measurement is ± 0.001 m or ± 1 mm. Measurements of 2 people won't agree exactly because each person's vision may be different. You should expect the tape measure to have less uncertainty because it probably covers the entire length of the room without having to be moved. The meter stick must be moved and replaced, creating an additional uncertainty when it is replaced each time.

2.11 Writing 4.8300 g implies that the balance measures to ±0.0001 g, when the correct measurement implies that it only measures to ±0.001 g.

2.12 Since the average is 41.18 mL, all the data are within ±0.1 mL of the average, and the first decimal place is uncertain by ±0.1 mL. Thus three significant figures should be used in reporting the measured volume.

2.13 a. Counting is an example of an exact number.

b. A measured quantity is not an example of an exact number.

c. A definition like 3 ft = 1 yd is an example of an exact number.

d. A measured quantity is not an example of an exact number.

2.14 The third decimal place is unknown in both 66.45 and 61.35, so the third decimal place in 5.100 cannot be given. See the rule for addition and subtraction.

2.15 The "12" is a defined number and does not limit the number of significant figures in the answer. Thus the answer is 3.083 or 3.08, with the same number of sig. figs. as 37.0 inches.

2.16 You should round only after the last step since you may lose some information or incur small errors from rounding up or down.

2.17 The factor = grain/g = 1 g/15.43 grain = 0.064808 = 0.06481 g/grain.

2.18 The factor = mg/μg = (1 mg/10^{-3}g) x (10^{-6} g/1 μg) = 10^{-3} mg/μg.

2.19 The density of a substance is the mass of the substance per unit volume, usually expressed as g/cm^3. The specific gravity of a substance is the ratio of the density of a substance (at any temperature) to the density of water at 4°C. Specific gravity has no units.

2.20 Gypsum will float in methylene iodide because it has a smaller density and is therefore lighter, and will sink in carbon tetrachloride because it has a larger density.

MEASUREMENT IN CHEMISTRY 9

■ Solutions to Practice Problems

Note on significant figures: Starting with problem 2.37, when the final answer is written for the first time, it is written first with one nonsignificant figure. The least significant digit is also underlined. The final answer is then rounded to the correct number of significant figures but the least significant digit is no longer underlined.

2.21 In converting a number in normal form to scientific notation, you should move the decimal point so as to obtain a number that is between 1 and less than 10. The number of places you move the decimal point gives the exponent. In parts b and d, note that the beginning zeroes are not significant, and do not appear in the answer. In part c, the terminal zeroes are significant, and do appear in the answer. The answers to each part are:

a. 3.0402×10^4 b. 6.58×10^{-3} c. 2.0900×10^2
d. 3.5×10^{-6} e. 1.046×10^3

2.22 As in the previous problem, the number of places you move the decimal point gives the exponent. In parts a and d, note that the beginning zeroes are not significant, and do not appear in the answer. In part c, the terminal zeroes are significant, and do appear in the answer. The answers to each part are:

a. 6.7×10^{-6} b. 6.705×10^2 c. 7.8000×10^1
d. 8.19×10^{-3} e. 5.901×10^3

2.23 Move the decimal point to the right if the exponent is positive; to the left if negative.

a. 890.45   b. 0.005126   c. 6.13
d. 0.00007302   e. 20,375.9

2.24 Move the decimal point to the right if the exponent is positive; to the left if negative.

a. 0.00631   b. 4.84   c. 256.89
d. 0.000033   e. 8,197.83

2.25 Follow the same directions as for the two problems above.

a. 26.8   b. 0.004598   c. 0.00783
d. 8,536.   e. 3,189.1

2.26 a. 0.000079823   b. 10,050.00   c. 0.0000872
d. 595,633.   e. 0.63

2.27 a. 3.69×10^{-3} m   b. 3.478×10^4 m   c. 9.2×10^{-11} s
d. 5.67×10^{-8} g

10 ■ CHAPTER 2

2.28 a. 4.53×10^{-1} g   b. 5.1×10^{-1} m   c. 1.03×10^{1} m
d. 4.57×10^{-12} g

2.29 a. 4.93 km   b. 2.3 μs   c. 5.68 mg
d. 1.568 cm   e. 9.3 ng

2.30 a. 6.15 ng   b. 7.48 cm   c. 3.9 mg
d. 4.37 μs   e. 8.2 km

2.31 In counting the number of significant figures, remember that the beginning zeroes are not significant, Thus in part a the zero to the left of the decimal point is not significant. Terminal zeroes are significant only where there is a decimal point. Thus the terminal zeroes in parts b, c, and e are significant, but the terminal zero in part d is not significant:

a. 4 sig. figs.   b. 3 sig. figs.   c. 4 sig. figs.
d. at least 2 sig. figs.   e. 3 sig. figs.

2.32 In counting the number of significant figures, remember that the beginning zeroes are not significant, Thus in parts d and e, the zero to the left of the decimal point is not significant. Terminal zeroes are significant only where there is a decimal point. Thus the terminal zeroes in parts b,c, and e are significant, but the terminal zero in part a is not significant:

a. 1 sig. fig.   b. 4 sig. figs.   c. 4 sig. figs.
d. 2 sig. figs.   e. 2 sig. figs.

2.33 The 0.00800 of part c has 3 sig figs because the 3 beginning zeroes are not significant.

2.34 The 0.4589 of part e has 4 sig figs because the zero to the left of the decimal point is a beginning zero and is not significant.

2.35 Round answers down when the first digit to be dropped is less than 5 and up when the it more than 5.

a. 308.0   b. 45.15   c. 126.6
d. 50.5092   e. 1.29379

2.36 Round answers down when the first digit to be dropped is less than 5 and up when the it more than 5.

a. 0.00526   b. 1.255   c. 49.667
d. 816.1   e. 816.0

MEASUREMENT IN CHEMISTRY ■ 11

2.37 As noted at the top of the practice problem section, the answer is given first with one nonsignificant figure and the least significant figure underlined. Then the answer is rounded off:

a. $561.25 + 68.499 = 6.297\underline{4}9 \times 10^2 = 6.2975 \times 10^2$

b. $75.23 - 70.451 = 4.7\underline{7}9 = 4.78$ (scientific notation not needed here)

c. $94.2 \times 2.456 = 2.3\underline{1}3 \times 10^2 = 2.31 \times 10^2$

d. $2.1 \times 44.56 = 9.\underline{3}57 \times 10^1 = 9.4 \times 10^1$

e. $5.897 \div 7.8 = 7.\underline{5}6 \times 10^{-1} = 7.6 \times 10^{-1}$

2.38 As noted at the top of the practice problem section, the answer is given first with one nonsignificant figure and the least significant figure underlined. Then the answer is rounded off:

a. $26 - 23.1 = \underline{2}.9 = 3.$ (scientific notation not needed here)

b. $687.2 + 5.64 = 692.\underline{8}4 = 6.928 \times 10^2$

c. $989.123 - 898.4981 = 90.62\underline{4}9 = 9.0625 \times 10^1$

d. $39.45 \times 19.8 = 78\underline{1}.11 = 7.81 \times 10^2$

e. $8.90 \div 563.29 = 0.0158\underline{0} = 1.58 \times 10^{-2}$

2.39 Follow the general procedures as outlined for problems above.

a. $(87.12 - 9.236) \div 5.489 = 14.1\underline{8}9 = 1.419 \times 10^1$

b. $65.456 \times (23.698 - 20.613) = 201.\underline{9}3 = 2.019 \times 10^2$

c. $(583.21 - 502.13 + 10.23) \div 90.135 = 1.01\underline{3}03 = 1.013$

d. $(73.54 \div 23.6) - 3.10 = 0.0\underline{1}61 = 2. \times 10^{-2}$

e. $(34.82 \times 6.4 \times 0.1439) - 2.68 = 2\underline{9}.38 = 2.9 \times 10^1$

12 ■ CHAPTER 2

2.40 a. $(26.5 \times 56.98) \div 67.19 = 22.\underline{4}7 = 2.25 \times 10^1$

b. $89.23 \times (56.23 - 51.65) = 40\underline{8}.6 = 4.09 \times 10^2$

c. $4.981 \div (9.105 - 8.987) = 42.\underline{2}1 = 4.22 \times 10^1$

d. $(89.1 \times 0.1256) - 3.51 = 7.6\underline{8}09 = 7.68$

e. $0.9548 + (5.16 \div 43.26) = 1.07\underline{4}0 = 1.074$

2.41 a. $165 \text{ L} \times 1 \text{ mL}/10^{-3} \text{ L} = 1.65 \times 10^5 \text{ mL}$

b. $48 \text{ ng} \times 10^{-9} \text{g}/1 \text{ ng} = 4.8 \times 10^{-8} \text{ g}$

c. $4.7 \text{ s} \times 1 \text{ ms}/10^{-3} \text{ s} = 4.7 \times 10^3 \text{ s}$

2.42 a. $23.1 \text{ g} \times 10^{-3} \text{ mg}/1 \text{ g} = 2.31 \times 10^4 \text{ mg}$

b. $2.15 \text{ ns} \times 10^{-9} \text{s}/1 \text{ ns} = 2.15 \times 10^{-9} \text{ s}$

c. $750. \text{ L} \times 1 \text{ mL}/10^{-3} \text{ L} = 7.50 \times 10^5 \text{ mL}$

2.43 a. $569 \text{ cm} \times (10^{-2} \text{ m}/1 \text{ cm}) \times (1 \text{ km}/10^3 \text{ m}) = 5.69 \times 10^{-3} \text{ km}$

b. $12.5 \text{ cm} \times (10^{-2} \text{ m}/1 \text{ cm}) \times (1 \text{ dm}/10^{-1} \text{ m}) = 1.25 \text{ dm}$

2.44 a. $87 \text{ ms} \times (10^{-3} \text{s}/1 \text{ ms}) \times (1 \text{ ns}/10^{-9} \text{ s}) = 8.7 \times 10^7 \text{ ns}$

b. $250.0 \text{ kg} \times (10^3 \text{ g}/1 \text{ kg}) \times (1 \text{ mg}/10^{-3} \text{ g}) = 2.500 \times 10^8 \text{ mg}$

2.45 $0.77 \text{ Å} \times \dfrac{100 \text{ pm}}{1 \text{ Å}} \times \dfrac{10^{-12} \text{ m}}{1 \text{ pm}} \times \dfrac{10^3 \text{ mm}}{1 \text{ m}} = 7.\underline{7}0 \times 10^{-8} = 7.7 \times 10^{-8} \text{ mm}$

2.46 $4.18 \times 10^{12} \text{ erg} \times \dfrac{10^{-7} \text{ J}}{1 \text{ erg}} \times \dfrac{1 \text{ kJ}}{10^3 \text{ J}} = 4.1\underline{8}0 \times 10^2 = 4.18 \times 10^2 \text{ kJ}$

2.47 $176 \text{ lb} \times 0.4536 \text{ kg}/1 \text{ lb} = 79.\underline{8}3 = 79.8 \text{ kg}$

2.48 $5.0 \text{ mi} \times 1.609 \text{ km}/1 \text{ mi} = 8.\underline{0}45 = 8.0 \text{ km}$

MEASUREMENT

2.49 40 km/hr x 1 mi/1.609 km = 2<u>4</u>.86 = 25 mi/hr

2.50 750 mL x 1L/10³ mL x 1 qt/0.9464 L = 0.79<u>2</u>47 = 0.7...

2.51 a. 329 K   b. 49°C   c. 308 K   d. 1...

2.52 a. 398 K   b. -85°C   c. 196 K   d. 122°C

2.53 273 + 30 = 303 K

2.54 77 - 273 = -196°C

2.55 No significant decimal places appear in any of the given temperatures, so the answers will not have any significant decimal places either. All digits to the left of the decimal point are significant in reporting temperatures.

   a. (85 - 32) ÷ 1.8 = 2<u>9</u>.4 = 29°C (note that the ".4" is not significant)

   b. (1.8 x -40) + 32 = -4<u>0</u>.0 = -40.°F (this temperature is the same on both scales)

   c. (-10 - 32) ÷ 1.8 = -2<u>3</u>.3 = 23°C (note that the ".3" is not significant)

   d. (1.8 x 55) + 32 = 13<u>1</u>.0 = -131°F (note that the ".0" is not significant)

   e. (1.8 x 125) + 32 = 25<u>7</u>.0 = 257°F (note that the ".0" is not significant)

2.56 No significant decimal places appear in any of the given temperatures, so the answers will not have any significant decimal places either. All digits to the left of the decimal point are significant in reporting temperatures.

   a. (1.8 x -15) + 32 = <u>5</u>.0 = 5°F (note that the ".0" is not significant)

   b. (75 - 32) ÷ 1.8 = 2<u>3</u>.9 = 24°C (note that the ".9" is not significant)

   c. (25 - 32) ÷ 1.8 = -<u>3</u>.9 = -4°C (note that the ".9" is not significant)

   d. (1.8 x 45) + 32 = 1<u>1</u>3.0 = 113°F (note that the ".0" is not significant)

   e. (1.8 x -75) + 32 = -10<u>3</u>.0 = -103°F (note that the ".0" is not significant)

2.57 $t_°F = (t_°C \times 1.8) + 32$ = (-78 x 1.8) + 32 = -10<u>8</u>.4 = -108

2.58 $t_°C = (t_°F - 32) \div 1.8$ = (-6 - 32) ÷ 1.8 = -2<u>1</u>.1 = -21.

2.59 Density = mass ÷ volume = 11.2 g ÷ 7.06 cm³ = 1.5<u>86</u> = 1.59 g/cm³

CHAPTER 2

Density = mass ÷ volume = 31.6 g ÷ 25.0 mL = 1.2<u>6</u>4 = 1.26 g/cm³

Density = mass ÷ volume = 0.975 g ÷ (5.97 - 5.60) mL = 2.<u>6</u>3 = 2.6 g/cm³

2.62   Density = mass ÷ volume = 32.6 g ÷ (28.7 - 25.0) mL = 8.<u>8</u>1 = 8.8 g/cm³

2.63   Density = mass ÷ volume; rearranging this definition to solve for volume gives:

volume = mass ÷ density = 24.5 g ÷ 1.049 g/mL = 23.<u>3</u>55 = 23.4 mL

2.64   Density = mass ÷ volume; rearranging this definition to solve for volume gives:

volume = mass ÷ density = 28.0 g ÷ 0.713 g/mL = 39.<u>2</u>7 = 39.3 mL

2.65   Density = mass ÷ volume; rearranging this definition to solve for mass gives:

mass = density x volume = 0.7908 g/mL x 24.48 mL = 19.3<u>5</u>8 = 19.36 g

2.66   Density = mass ÷ volume; rearranging this definition to solve for mass gives:

mass = density x volume = 0.875 g/mL x 56.7 mL = 49.<u>6</u>1 = 49.6 g

2.67   Density = mass ÷ volume; rearranging and using density = specific gravity gives:

mass = density x volume = 1.250 g/mL x 1.20 x 10³ mL = 15<u>0</u>0
= 1.50 x 10³ g (1.50 kg)

2.68   Density = mass ÷ volume; rearranging and using density = specific gravity gives:

volume = mass ÷ density = 408 g ÷ 0.863 g/mL = 47<u>2</u>.7 = 4.73 x 10² mL (0.473 L)

■ Solutions to Additional Problems

2.69   Since the average is 10.0536 mL, all the data are within ±0.01 mL of the average, and the second decimal place is uncertain by ±0.01 mL. Thus four significant figures should be used in reporting the measured volume.

2.70   The average for the first student is 29.57 g, and the average for the second is 29.5 g.

MEASUREMENT IN CHEMISTRY ■ 15

2.71 Remember that terminal zeroes are significant when there is a decimal point.

    a. 9 sig. figs.   b. 3 sig. figs.   c. 2 sig. figs.
    d. 3 sig. figs.   e. 3 sig. figs.

2.72 Beginning zeroes are not significant; terminal zeroes are significant if there is a decimal point.

    a. 8 sig. figs.   b. 3 sig. figs.   c. 2 sig. figs.
    d. 4 sig. figs.   e. 3 sig. figs.

2.73   a. 6.41   b. 1.5×10^1 c. 8.30×10^{-1}
         d. 6.07   e. 2.479×10^2

2.74   a. 9.19×10^1 b. 8.94×10^2 c. 8.30
         d. 1.78×10^1 e. 2.19×10^1

2.75 2×10^{12} molecules

2.76   0.0000035 m

2.77   a. 4.37×10^{-5} b. 5.720×10^5 c. 1.0×10^{-5}
         d. 1.5×10^{-5} e. 1.61×10^{-2}

2.78   a. 1.463×10^5 b. -2.4×10^{-3} c. 1.32
         d. 1.781×10^{-3} e. -2.6

2.79   a. 5.489 km   b. 723 ng   c. 2.164×10^1 km
         d. 6.50×10^2 μL   e. 2.3 ms

2.80   a. 2.47 mg   b. 3.5 cm or 35 mm   c. 5 μs
         d. 6.815 km   e. 5.214×10^2 pg

2.81 $(1.8 \times 113°) + 32 = 235.4 = 235°F$; $113 + 273 = 386$ K

2.82 $(1.8 \times -7°) + 32 = 19.4 = 19°F$; $-7 + 273 = 266$ K

2.83 $(93° - 32°) \div 1.8 = 33.8 = 34°C$; $34 + 273 = 307$ K

2.84 $(23° - 32°) \div 1.8 = -5.0 = -5°C$; $-5 + 273 = 268$ K

2.85   Volume = mass ÷ density = 0.0456 g ÷ 1.231 g/mL = 0.03704 = 3.70×10^{-2} mL

2.86   Volume = mass ÷ density = 0.324 g ÷ 1.35 g/cm^3 = 0.2400 = 2.40×10^{-1} cm^3

2.87   Mass = density × volume = 1.063 g/cm^3 × 2.35 cm^3 = 2.498 = 2.50 g

2.88 Mass = density x volume = 1.093 g/cm^3 x 5.67 cm^3 = 6.1<u>9</u>7 = 6.20 g

2.89 Use a density equal to the specific gravity, and equate the volume to 1.50 x 10^3 mL:

Mass = density x volume = 1.28 g/mL x 1.50 X 10^3 mL = 1.9<u>2</u>0 x 10^3 g or 1.92 kg

2.90 Use a density equal to the specific gravity, and equate the volume to 2.50 x 10^3 mL:

Mass = density x volume = 1.048 g/mL x 2.50 X 10^3 mL = 2.6<u>2</u>0 x 10^3 g or 2.62 kg

2.91 0.5 mm x $\dfrac{10^{-3} \text{ m}}{1 \text{ mm}}$ x $\dfrac{1 \text{ μm}}{10^{-6} \text{ m}}$ = <u>5</u>.0 x 10^2 = 5 x 10^2 μm

2.92 2 μm x $\dfrac{10^{-6} \text{ m}}{1 \text{ μm}}$ x $\dfrac{1 \text{ mm}}{10^{-3} \text{ m}}$ = <u>2</u>.0 x 10^{-3} = 2 x 10^{-3} mm

2.93 6 lb x 0.4536 kg/1 lb = <u>2</u>.72 = 3 kg

2.94 11 km x $\dfrac{1 \text{ mi}}{1.609 \text{ km}}$ x $\dfrac{5280 \text{ ft}}{1 \text{ mi}}$ = 3.<u>6</u>09 x 10^4 = 3.6 x 10^4 ft

2.95 60 μm x $\dfrac{10^{-6} \text{ m}}{1 \text{ μm}}$ x $\dfrac{1 \text{ cm}}{10^{-2} \text{ m}}$ x $\dfrac{1 \text{ in}}{2.54 \text{ cm}}$ = 2.<u>3</u>6 x 10^{-3} = 2.4 x 10^{-3} in

2.96 6684 ft x $\dfrac{1 \text{ yd}}{3 \text{ ft}}$ x $\dfrac{0.9144 \text{ m}}{1 \text{ yd}}$ = 2.03<u>7</u>3 x 10^3 = 2.037 x 10^3 m

3. MATTER AND ENERGY

Solutions to Exercises

Note on significant figures: Starting with Exercise 3.3, the first time the final answer is written, it is written first with one nonsignificant figure. The least significant digit is also underlined. The final answer is then rounded to the correct number of significant figures but the least significant digit is no longer underlined.

3.1 It is a solid.

3.2 Its physical properties are: colorless, physical state is a liquid, boiling point = 78°C, and freezing point = -114°C. Its only listed chemical property is that it burns forming CO_2 + H_2O.

3.3 Using the conversion factor of 1 cal/4.184 J:

48.6 J x 1 cal/4.184 J = 11.6̲1 = 11.6 cal

3.4 Heat = 42 g x $\dfrac{0.129 \text{ J}}{1 \text{ g} \cdot 1°C}$ x (100°C - 26°C) = 4.0̲1 x 10^2 = 4.0 x 10^2 J

3.5 sp heat = 5.68 J x $\dfrac{1}{8.75 \text{ g}}$ x $\dfrac{1}{(30.35°C - 25.47°C)}$

= 1.33̲02 x 10^{-1} = 1.33 x 10^{-1} J/(g·°C)

Answers to Questions to Test Your Reading

3.1 Three examples are: 1) water in the liquid form is changed by freezing to ice, 2) water in liquid form is boiled and forms steam, and 3) water as ice is melted to form liquid water.

3.2 Paper (and possibly wood) is a material that exists as a solid but not a gas or liquid.

3.3 The name is the melting point (it usually equals the freezing point).

3.4 The term vapor refers to the gaseous state of a substance that normally exists as a liquid or solid.

3.5 The pressure would have to be decreased to lower the boiling point.

3.6 Three examples are: 1) sawing a log into boards, 2) changing liquid water into ice, and 3) dissolving solid salt in water.

3.7 Three examples are: 1) burning a log to form smoke and ashes, 2) rusting iron, and 3) reacting mercury metal with oxygen to form mercury(II) oxide.

3.8 Three examples are: 1) melting point, 2) boiling point, and 3) color.

3.9 Iron: reacts with oxygen to form rust; mercury: reacts with oxygen to form mercury(II) oxide; and bromine: reacts with sodium to form a white solid.

3.10 A substance is a material that cannot be separated into different materials by any physical process. An example is liquid water.

3.11 A mixture is a material that can be separated into two or more substances. An example is a mixture of salt and water.

3.12 Filtration depends on pouring a mixture such as coffee grounds and coffee extract through a paper filter cup or cone. The coffee extract passes through the filter, leaving the grounds behind in the paper filter.

3.13 When a solution of salt in water is heated to the boiling point, the water vapor passes out of the distillation flask, into the condenser where it cools and runs out into the receiver flask.

3.14 Two examples are: 1) a liquid solution of sodium chloride in water, and 2) air, which is a gaseous mixture of nitrogen, oxygen, etc.

3.15 Two examples are: 1) a mixture of sugar and salt crystals, and 2) sand, which is a mixture of calcite, silica, etc.

MATTER AND ENERGY ■ 19

3.16 A phase is a homogeneous part of any given portion of a material. A glass containing liquid water with an ice cube floating on top consists of two phases of water: liquid water and ice.

3.17 An element is a substance that cannot be decomposed by any chemical reaction into simpler substances. Three examples are: 1) mercury metal, 2) oxygen gas, and 3) iron metal.

3.18 A compound is a substance composed of two or more elements chemically combined. Three examples are: 1) sugar, 2) salt, and 3) aspirin.

3.19 When a compound like water is separated into its elements (for example by passing an electric current through it), the elements are formed in definite or constant proportions by mass.

3.20 Energy is the potential or capacity to move matter. An example of kinetic energy is the electric current from a battery going to an electric motor which moves an electric car; the battery has the capacity to move matter. An example of potential energy is the same battery in an electric car at rest; the battery has the potential to move matter.

3.21 If the distance is expressed in meters, then multiplying $\underline{F} \times \underline{d}$ gives units of kg · m²/s², which are the SI units for joules:

$$\frac{kg \cdot m}{s^2} \times m = \frac{kg \cdot m^2}{s^2} = J$$

3.22 The specific heat of a substance is the quantity of heat required to raise the temperature of one gram of that substance by one degree Celsius. In SI units, the specific heat of water is 4.18 J/(g · °C)

3.23 The specific heat is 5.5 J/(g · °C).

3.24 Once the pendulum is moved upward and held at some point higher than its than the lowest point, it is at rest with all of its energy in the form of potential energy. After it is released, some of the potential energy is converted to kinetic energy, but the total of both energies is equal to the amount of the potential energy at the start. Thus energy is conserved.

Solutions to Practice Problems

Note on significant figures: Starting with problem 35, the first time the final answer is written, it is written first with one nonsignificant figure. The least significant digit is also underlined. The final answer is then rounded to the correct number of significant figures but the least significant digit is no longer underlined.

3.25   It is a liquid.

3.26   It is a gas.

3.27   It is a gas.

3.28   It is a liquid.

3.29   Its physical properties are: colorless, 69°C boiling point, and -95°C freezing point. Its chemical property is that it burns to form CO_2 and H_2O

3.30   Its physical properties are: blue-black as a solid, violet as a gas, 114°C melting point, and 185°C boiling point. Its chemical property is that it reacts with sodium to form a white solid.

3.31   Its physical properties are: colorless, oily liquid, and density of 1.59 g/cm³. Its chemical properties are: dilates blood vessels, and explodes forming N_2, CO_2, H_2O, and O_2.

3.32   Its physical properties are: white solid, and sweet (but poisonous) taste. Its chemical property is that it reacts with NaI to form yellow PbI_2.

3.33   a. mixture   b. element   c. compound   d. mixture

3.34   a. mixture   b. element   c. compound   d. mixture

3.35   Using 121 kJ = 1.21×10^5 J, and the conversion factor of 1 cal/4.184 J:

   1.21×10^5 J x 1 cal/4.184 J = $2.8\underline{9}1 \times 10^4$ = 2.89×10^4 cal

3.36   Using the conversion factor of 1 cal/4.184 J:

   321 J x 1 cal/4.184 J = $7.6\underline{7}2 \times 10^1$ = 7.67×10^1 cal

3.37   Using the conversion factor of 4.184 J/1 cal:

   12.0×10^3 cal x 4.184 J/1 cal = $5.0\underline{2}08 \times 10^4$ = 5.02×10^4 J

MATTER AND ENERGY ■ 21

3.38 Using the conversion factor of 4.184 J/1 cal:

181 cal × 4.184 J/1 cal = 7.5$\underline{7}$3 × 10² = 7.57 × 10² J

3.39 Heat = 235 g × $\dfrac{0.901 \text{ J}}{1\text{g} \cdot 1°\text{C}}$ × (15.0°C) = 3.1$\underline{7}$6 × 10³ = 3.18 × 10³ J

3.40 Heat = 35.8 g × $\dfrac{0.384 \text{ J}}{1\text{g} \cdot 1°\text{C}}$ × (25.0°C) = 3.4$\underline{3}$6 × 10² = 3.44 × 10² J

3.41 Heat = 65.4 g × $\dfrac{0.741 \text{ J}}{1\text{g} \cdot 1°\text{C}}$ × (38.5 − 20.0)°C = 8.9$\underline{6}$53 × 10² = 8.97 × 10² J

3.42 Heat = 500 g × $\dfrac{0.649 \text{ J}}{1\text{g} \cdot 1°\text{C}}$ × (45.0 − 20.0)°C = 8.1$\underline{1}$2 × 10³ = 8.11 × 10³ J

3.43 sp heat = 151.5 J × $\dfrac{1}{18.9 \text{ g}}$ × $\dfrac{1}{(39.9°\text{C} - 23.2°\text{C})}$

= 4.7$\underline{9}$9 × 10⁻¹ = 4.80 × 10⁻¹ J/(g·°C)

3.44 sp heat = 189.5 J × $\dfrac{1}{28.3 \text{ g}}$ × $\dfrac{1}{(55.7°\text{C} - 24.8°\text{C})}$

= 2.1$\underline{6}$7 × 10⁻¹ = 2.17 × 10⁻¹ J/(g·°C)

3.45 Increase, °C = 1071 J × $\dfrac{1}{36.0 \text{ g}}$ × $\dfrac{1\text{g} \cdot 1°\text{C}}{2.43 \text{ J}}$ = 12.$\underline{2}$4 = 12.2 °C

3.46 Increase, °C = 1628 J × $\dfrac{1}{48.6 \text{ g}}$ × $\dfrac{1\text{g} \cdot 1°\text{C}}{2.51 \text{ J}}$ = 13.$\underline{3}$4 = 13.3 °C

3.47 Increase, °C = 164.8 J × $\dfrac{1}{15.5 \text{ g}}$ × $\dfrac{1\text{g} \cdot 1°\text{C}}{0.449 \text{ J}}$ = 23.$\underline{6}$7 °C

Final temp = 21.5°C + 23.$\underline{6}$7°C = 45.$\underline{1}$7 = 45.2°C

3.48 Increase, °C = 104 J × $\frac{1}{16.8 \text{ g}}$ × $\frac{1 \text{g} \cdot 1°\text{C}}{0.444 \text{ J}}$ = 13.9̲4 °C

Final temp = 25.2°C + 13.9̲4°C = 39.1̲4 = 39.1°C

■ Solutions to Additional Problems

3.49	a. Gas	b. Liquid	c. Gas	d. Liquid	e. Solid
3.50	a. Solid	b. Liquid	c. Solid	d. Gas	e. Gas
3.51	a. Physical	b. Chemical	c. Physical	d. Physical	e. Chemical
3.52	a. Physical	b. Chemical	c. Physical	d. Physical	e. Chemical
3.53	a. Solid	b. Solid	c. Solid	d. Solid	e. Gas
3.54	a. Solid	b. Liquid	c. Gas	d. Gas	e. Gas
3.55	a. Solid	b. Liquid	c. Liquid	d. Gas	e. Gas
3.56	a. Liquid	b. Liquid	c. Liquid	d. Gas	e. Gas
3.57	a. Substance	b. Het. mixt.	c. Solution	d. Substance	
3.58	a. Substance	b. Substance	c. Solution	d. Het. mixt.	
3.59	a. Physical	b. Chemical	c. Physical	d. Chemical	
3.60	a. Chemical	b. Chemical	c. Physical	d. Chemical	

3.61 453 Btu × 252 cal/Btu = 1.14̲1 × 10^5 = 1.14 × 10^5 cal

1.141 × 10^5 cal × 4.184 J/1 cal = 4.7̲73 × 10^5 = 4.77 × 10^5 J

3.62 153 kwh × 3.60 × 10^6 J/kwh = 5.50̲8 × 10^8 = 5.51 × 10^8 J

5.508 × 10^8 J × 1 cal/4.184 J = 1.31̲6 × 10^8 = 1.32 × 10^8 cal

3.63 3.5 × 10^4 cal × $\frac{1\text{g} \cdot 1°\text{C}}{1 \text{ cal}}$ × $\frac{1}{(35.0 - 20.0)°\text{C}}$ = 2.3̲3 × 10^3 = 2.3 × 10^3 g

3.64 $4.0 \times 10^4 \text{ cal} \times \dfrac{1 g \cdot 1°C}{1 \text{ cal}} \times \dfrac{1}{(40.0 - 25.0)°C} = 2.6\underline{6} \times 10^3 = 2.7 \times 10^3 \text{ g}$

3.65 $\dfrac{4.18 \text{ J}}{1 g \cdot 1°C} \times 5.42 \text{ g} \times (45.1 - 20.8)°C = 5.5\underline{0}53 \times 10^2 = 5.51 \times 10^2 \text{ J}$

3.66 $\dfrac{2.06 \text{ J}}{1 g \cdot 1°C} \times 6.87 \text{ g} \times [-1.5 - (-15.0)]°C = 1.9\underline{1}05 \times 10^2 = 1.91 \times 10^2 \text{ J}$

4. ATOMS, MOLECULES, AND IONS

■ Solutions to Exercises

<u>Note on significant figures:</u> Starting with Exercise 4.4, the first time the final answer is written, it is written first with one nonsignificant figure. The least significant digit is also underlined. The final answer is then rounded to the correct number of significant figures but the least significant digit is no longer underlined.

4.1   The atomic number of the atom is 17 (number of protons). The mass number is 17 + 18 = 35. From the list of elements on the inside back cover, we see that the symbol of the element with atomic number 17 is chlorine. The isotope symbol is:

$$^{35}_{17}Cl$$

4.2   The left subscript of 9 gives the number of protons for fluorine. The left superscript gives the mass number, so the number of neutrons in the nucleus is

number of neutrons = mass number - atomic number
number of neutrons = 19 - 9 = 10

4.3   From the inside back cover, we see that boron has an atomic number of 5, indicating 5 protons and 5 electrons. The number of neutrons = mass number - atomic number; so 11 - 5 = 6 neutrons. The protons and neutrons are located in the nucleus whereas the electrons are located around and far away from the nucleus.

4.4   To find the atomic weight of copper, multiply each isotopic mass by its fractional abundance. Then add the products to obtain the atomic weight:

62.94 amu x 0.6917 = 43.5<u>3</u>6 amu
64.93 amu x 0.3083 = 20.0<u>1</u>8 amu
----
atomic weight   = 63.5<u>5</u>4 = 63.55 amu

ATOMS, MOLECULES, AND IONS ■ 25

4.5 The element in Group IIA, period 3 is magnesium. The symbol of the element is Mg.

4.6 Using the periodic table (Fig. 4.10), we see that nitrogen (N) and phosphorus (P) are the nonmetallic elements in Group VA.

4.7 Methane is composed to carbon and hydrogen, both nonmetals, indicating that methane is molecular. Because it is a gas at room temperature, its melting point is below room temperature, which is expected of a molecular substance.

4.8 This molecule consists of phosphorus and oxygen so its molecular formula will contain their respective symbols P and O. Since there are four atoms of phosphorus and ten oxygen atoms, the formula is P_4O_{10}.

4.9 Barium is a metal, so we expect a cation. Because barium is in Group IIA, the group number is 2 and the expected cation charge is +2.

■ Answers to Questions to Test Your Reading

4.1 According to Dalton, an <u>element</u> is a type of matter composed of only one kind of atom with each of those atoms having the same properties. He defined a <u>compound</u> as a type of matter composed of atoms of two or more elements chemically combined in fixed proportions. He explained a <u>chemical reaction</u> as a rearrangement of the atoms present in the reacting substance, forming new chemical combinations in the products.

4.2 Atomic theory states that because the number of each kind of atom is the same before and after a reaction, the total mass remains constant. Atomic theory says that because a compound contains atoms in definite, fixed proportions, and because atoms have define mass, the elements in a compound must be present in definite proportion by mass.

4.3 Two electric charges of opposite kinds will attract each other, so the cork balls will attract each other.

4.4 Cathode rays are produced in an electric discharge tube containing a cathode and an anode. When high voltage is fed to the cathode, the cathode emits a cathode ray or stream of electrons, which are attracted to the positive anode.

4.5 Geiger and Marsden found that most alpha particles passed through the metal foils, with a few being deflected nearly backward. Rutherford concluded that atoms in the foils had a positively charge nuclear core that takes up very little space but contains most of the mass.

26 ■ CHAPTER 4

4.6 The one- or two-letter symbol identifies the element. The left subscript gives the atomic number as well as the number of protons and electrons. The left superscript gives the atomic mass or total of the number of protons and neutrons. The number of neutrons is obtained by subtracting the subscript from the superscript.

4.7 Inside the sodium nucleus are 11 protons and 12 neutrons; outside the sodium nucleus are 11 electrons.

4.8 The percentage abundances are generally the same because for naturally occurring elements the percentages have remained constant over time and in most cases are independent of the origin of the element.

4.9 The carbon-12 isotope is the modern standard used for obtaining other masses.

4.10 Modern periodic tables differ in the respect that they are arranged by increasing atomic numbers, not the atomic weights used by Mendeleev.

4.11 The letter "B" indicates that copper is a transition-metal element.

4.12 The Group IA metals are similar in that they are very reactive chemically, combing with most nonmetals. Their physical properties also change in a regular fashion.

4.13 The two rows at the bottom are not connected because it is difficult to include a row of 32 elements, with all the usual information, on a text page.

4.14 A metalloid, like silicon, has properties intermediate between those of a metal and those of a nonmetal. It is also a semiconductor of electricity.

4.15 The two main types of compounds are molecular compounds (example: water) and ionic compounds (example: sodium chloride). Ionic compounds tend to melt at much higher temperatures whereas many molecular compounds are liquid or gaseous.

4.16 A <u>molecular formula</u> is a notation that uses atomic symbols with subscripts to indicate the total number of atoms of different elements in one molecule the substance (example: H_2O). A simple <u>structural formula</u> shows how much each atom is bonded to other atoms but does not "total up" the number of atoms in the compound (example: H-O-H).

4.17 Sodium chloride has a <u>crystalline</u> structure with varying but equal numbers of Na^+ and Cl^- ions in the crystal. Each Na^+ ion is surrounded by 6 Cl^- ions, and vice versa. The crystal is electrically neutral.

4.18 It is not written as $Fe_2S_3O_{12}$ because the anion structure is that of three SO_4^{2-} ions. Using $S_3O_{12}^{6-}$ implies that the structure of the anion involves 12 oxygens bonded to three sulfurs in just one anion.

Solutions to Practice Problems

Note on significant figures: Starting with problem 4.31, the first time the final answer is written, it is written first with one nonsignificant figure. The least significant digit is also underlined. The final answer is then rounded to the correct number of significant figures but the least significant digit is no longer underlined.

4.19   a. $^{20}_{10}Ne$ b. $^{28}_{14}Si$ c. $^{31}_{15}P$ d. $^{39}_{19}K$ e. $^{55}_{25}Mn$

4.20   a. $^{13}_{6}C$ b. $^{24}_{12}Mg$ c. $^{40}_{18}Ar$ d. $^{50}_{24}Cr$ e. $^{58}_{28}Ni$

4.21   a. 7p + 7n   b. 15p + 17n   c. 16p + 18n   d. 17p + 18n   e. 24p + 28 n

4.22   a. 10p + 12n   b. 14p + 15n   c. 20p + 20n   d. 27p + 32n   e. 29p + 34 n

4.23   Zinc-64 has 30 protons, 34 neutrons, and 30 electrons. The protons and neutrons are in the nucleus; the electrons are far outside the nucleus.

4.24   Argon-40 has 18 electrons and 22 neutrons.

4.25   Selenium-80 has 34 electrons.

4.26   Sulfur-32 has 16 electrons. Bromine-79 has 35 electrons.

4.27   Arsenic-75 has 33 protons and 42 neutrons in the nucleus and 33 electrons outside the nucleus.

4.28   Silicon-29 has 14 protons and 15 neutrons in the nucleus and 14 electrons outside the nucleus.

4.29   Nitrogen-14 is far more abundant.

4.30   Vanadium-51 is far more abundant.

4.31   To find the atomic weight of boron, multiply each isotopic mass by its fractional abundance. Then add the products to obtain the atomic weight:

$$10.01 \text{ amu} \times 0.197 = 1.9\underline{7}2 \text{ amu}$$
$$11.01 \text{ amu} \times 0.803 = 8.8\underline{4}1 \text{ amu}$$

atomic weight = 10.8$\underline{1}$3 = 10.81 amu

4.32 To find the atomic weight of copper, multiply each isotopic mass by its fractional abundance. Then add the products to obtain the atomic weight:

62.93 amu x 0.691 = 43.48 amu
64.93 amu x 0.309 = 20.06 amu
_____
atomic weight = 63.54 = 63.5 amu

4.33 To find the atomic weight of the unknown element, multiply each isotopic mass by its fractional abundance. Then add the products to obtain the atomic weight:

19.99 amu x 0.9092 = 18.175 amu
20.99 amu x 0.00257 = 0.05394 amu
21.99 amu x 0.0882 = 1.939 amu
_____
atomic weight = 20.1679 = 20.17 amu (The element is neon.)

4.34 To find the atomic weight of the unknown element, multiply each isotopic mass by its fractional abundance. Then add the products to obtain the atomic weight:

27.98 amu x 0.9218 = 25.791 amu
28.98 amu x 0.0471 = 1.365 amu
29.97 amu x 0.0312 = 0.9351 amu
_____
atomic weight = 28.091 = 28.09 amu (The element is silicon.)

4.35 a. In: indium   b. Si: silicon   c. Mn: manganese
d. Ti: titanium   e. As: arsenic

4.36 a. Mg: magnesium   b. K: potassium   c. S: sulfur
d. Au: gold   e. Sn: tin

4.37 Bismuth(Bi) is the only metal in Group VA.

4.38 Silicon and germanium are the only metalloids in Group IVA.

4.39 Phosphorus: atomic no. = 15; atomic wt = 30.97 or 30.973762, group VA, period 3, nonmetal.

4.40 Selenium: atomic no. = 34; atomic wt = 78.96, group VIA, period 4, nonmetal.

4.41   a. C + O form molecular compound(s)

b. N + O form molecular compound(s)

c. Li + O form ionic compound(s)

d. Ca + Cl form ionic compound(s)

e. S + Cl form molecular compound(s)

4.42   a. Na + N form ionic compound(s)

b. P + S form molecular compound(s)

c. H + S form molecular compound(s)

d. Ba + O form ionic compound(s)

e. Mg + Cl form ionic compound(s)

4.43   Sodium, sulfur, and oxygen should form an ionic compound of the type $Na_xS_yO_z$. where Na^+ is the cation and the $S_yO_z^{n-}$ is a polyatomic anion. This prediction agrees with the high melting point.

4.44   Carbon, hydrogen, and iodine should form a molecular compound of the type $C_xH_yI_z$. where carbon is bonded to both the other elements. This prediction agrees with the low melting point.

4.45   a. H_2SO_4 b. Cl_2O_7 c. N_2O_4
d. C_3H_8 e. N_2F_4

4.46   a. H_2S b. PCl_5 c. C_6H_6
d. HNO_3 e. $HClO_4$

4.47   a. +1   b. -1   c. -3   d. +2   e. +2

4.48   a. -1   b. +2   c. -2   d. +3   e. +1

4.49   Two Na^+ ions and one CO_3^{2-} ion.

4.50   Two Al^{3+} ions and three SO_4^{2-} ions.

Solutions to Additional Problems

4.51 Sodium-24 has 11 protons, 11 electrons, and 13 neutrons.

4.52 Technetium-99 has 43 protons, 43 electrons, and 56 neutrons.

4.53 The isotope symbol for this calcium isotope is:
$${}^{47}_{20}Ca$$

4.54 The isotope symbol for this iron isotope is:
$${}^{54}_{26}Fe$$

4.55 They differ by two neutrons; they are alike in that both have 28 protons and 28 electrons.

4.56 They differ by two neutrons; they are alike in that both have 30 protons and 30 electrons.

4.57 The proton is 1.8×10^3 times heavier than the electron.

4.58 The neutron is 1.00137 times as heavy as the proton.

4.59 To find the atomic weight of chromium, multiply each isotopic mass by its fractional abundance. Then add the products to obtain the atomic weight:

49.95 amu x 0.0431 = 2.153 amu
51.94 amu x 0.8376 = 43.505 amu
52.94 amu x 0.0955 = 5.056 amu
53.94 amu x 0.0238 = 1.283 amu

atomic weight = 51.997 = 52.00 amu

4.60 To find the atomic weight of iron, multiply each isotopic mass by its fractional abundance. Then add the products to obtain the atomic weight:

53.94 amu x 0.0584 = 3.1501 amu
55.94 amu x 0.9168 = 51.286 amu
56.94 amu x 0.0217 = 1.236 amu
57.93 amu x 0.0031 = 0.179 amu

atomic weight = 55.8511 = 55.85 amu

4.61 Group IVA: C (nonmetal); Si (metalloid); Ge (metalloid); Sn (metal); and Pb (metal). The trend in this group does agree with the general rule.

4.62 Period 2: Li (metal); Be (metal); B (metalloid); C (nonmetal); N (nonmetal); O (nonmetal); F (nonmetal); and Ne (nonmetal). The trend in this period agrees with the general rule.

4.63 The molecular formula for chloroform is $CHCl_3$.

4.64 The molecular formula for formaldehyde is CH_2O.

4.65 The SO_4^{2-} ion formula means that there is one sulfur atom in the ion and four oxygen atoms in the ion, and that the charge on the ion is -2.

4.66 The HCO_3^- ion formula means that there is one hydrogen atom in the ion, one carbon atom in the ion, and four oxygen atoms in the ion, and that the charge on the ion is -1.

5. CHEMICAL FORMULAS AND NAMES

■ Solutions to Exercises

5.1   a. Li^+ and F^- form LiF b. Li^+ and S^{2-} form Li_2S c. Li^+ and N^{3-} form Li_3N

5.2   a. K_2S, potassium sulfide   b. Li_3N, lithium nitride   c. Al_2O_3, aluminum oxide

5.3   Barium sulfide contains the barium ion, Ba^{2+}, and the sulfide ion, S^{2-}. The formula is BaS. Lithium chloride contains the lithium ion, Li^+, and the chloride ion, Cl^-. The formula is $LiCl$. Silver nitride contains the silver ion, Ag^+, and the nitride ion, N^{3-}. The formula is Ag_3N.

5.4 SnO contains the tin ion, Sn^{2+}, and the oxide ion O^{2-}. According to the Stock method, the name of the compound is tin(II) oxide. Its classical name is stannous oxide. SnO_2 contains the tin ion, Sn^{4+}, and the oxide ion, O^{2-}. According to the Stock method, the name of the compound is tin(IV) oxide. Its classical name is stannic oxide.

5.5   a. K_2SO_3, potassium sulfite   b. $Ba(ClO)_2$, barium hypochlorite
       c. NH_4CN, ammonium cyanide

5.6   Barium phosphate contains the barium ion, Ba^{2+}, and the phosphate ion, PO_4^{3-}. The formula of the compound is $Ba_3(PO_4)_2$.

5.7   a. N_2O_3, dinitrogen trioxide   b. ClF_3, chlorine trifluoride
       c. CCl_4, carbon tetrachloride

5.8   Hydrofluoric acid has the formula $HF(aq)$.

5.9   The oxyacid corresponding to the sulfite ion, SO_3^{2-}, is sulfurous acid. The formula is H_2SO_3.

Answers to Questions to Test Your Reading

5.1 A <u>binary ionic compound</u> is a compound that is composed of ions from only two elements. In most cases, these compounds consist of a metal, which is the cation, and a nonmetal, which is than anion. An example is sodium chloride, NaCl.

5.2 When naming a binary ionic compound, the cation is always named first.

5.3 Cations are named after the metallic element from which they are derived. Thus, the names of the ions are: sodium ion, potassium ion, magnesium ion, and calcium ion.

5.4 The names of the ions are fluoride, chloride, bromide, and iodide.

5.5 According to the Stock method, the names of the Cu^+ and Cu^{2+} ions are copper(I) and copper(II), respectively. Their classical names are cuprous ion and cupric ion, respectively.

5.6 According to the Stock method, the names of the Pb^{2+} and Pb^{4+} ions are lead(II) and lead(IV), respectively. Their common names are plumbous and plumbic, respectively.

5.7 A <u>polyatomic ion</u> is a group of chemically bonded atoms that function as a single unit. An <u>oxyanion</u> is a negatively charged polyatomic ion that contains an atom of some element in addition to one or more oxygen atoms. An example of a polyatomic ion that is not an oxyanion is NH_4^+. an example of a polyatomic ion that is an oxyanion is the carbonate ion, CO_3^{2-}.

5.8 A <u>binary molecular compound</u> is a compound that contains two chemically bonded nonmetal atoms forming molecules. It differs from a binary ionic compound in that it does not contain ions. Also, binary ionic compounds usually contain a metal whereas a binary molecular compound does not. Finally, binary ionic compounds do not form molecules.

5.9 An <u>acid</u> is a compound that produces hydrogen ion (H^+) when it is dissolved in water. A <u>binary acid</u> is a substance (dissolved in water) that contains hydrogen and another nonmetallic element. An example of a binary acid is hydrochloric acid, HCl(aq).

5.10 An <u>oxyacid</u> is a molecular substance containing hydrogen, oxygen, and another element. When added to water, this acid yields hydrogen (H^+) ions and oxyanions. An example is nitric acid, HNO_3.

Solutions to Practice Problems

5.11 Al_2O_3

5.12 Ca_3P_2

5.13
a. MgO
b. $CaCl_2$
c. Na_2O
d. AlF_3
e. CsBr
f. Zn_3N_2

5.14
a. CdS
b. KF
c. BaI_2
d. Na_3N
e. Ag_2S
f. Mg_3P_2

5.15
a. Li^+ and O^{2-}
b. Ba^{2+} and O^{2-}
c. Zn^{2+} and Cl^-
d. K^+ and P^{3-}
e. Na^+ and Br^-
f. Ca^{2+} and I^-

5.16
a. Cd^{2+} and I^-
b. Ca^{2+} and P^{3-}
c. Na^+ and S^{2-}
d. Mg^{2+} and S^{2-}
e. Li^+ and I^-
f. Mg^{2+} and F^-

5.17 sodium chloride

5.18 lithium chloride

5.19
a. lithium bromide
b. calcium bromide
c. rubidium oxide
d. barium nitride
e. zinc sulfide
f. aluminum fluoride

5.20
a. sodium iodide
b. cesium sulfide
c. cadmium oxide
d. aluminum chloride
e. magnesium phosphide
f. magnesium bromide

5.21
a. BaO
b. $ZnCl_2$
c. Na_2S
d. NaBr
e. Al_2S_3
f. K_3N

5.22
a. CaI_2
b. $BaBr_2$
c. Ag_2O
d. CdO
e. Na_3P
f. AlF_3

5.23
a. cobalt(III) oxide, cobaltic oxide
b. cobalt(II) oxide, cobaltous oxide
c. lead(IV) oxide, plumbic oxide
d. lead(II) oxide, plumbous oxide
e. copper(II) oxide, cupric oxide
f. copper(I) oxide, cuprous oxide

5.24
a. tin(IV) oxide, stannic oxide
b. tin(II) oxide, stannous oxide
c. iron(III) sulfate, ferric sulfate
d. iron(II) sulfate, ferrous sulfate
e. cobalt(III) fluoride, cobaltic fluoride
f. cobalt(II) fluoride, cobaltous fluoride

CHEMICAL FORMULAS AND NAMES ■ 35

5.25   a. SnCl$_4$ is tin(IV) chloride, or stannic chloride.
       b. correct
       c. CuBr$_2$ is copper(II) bromide, or cupric bromide.
       d. Fe$_2$O$_3$ is iron(III) oxide, or ferric oxide.
       e. correct
       f. CaCl$_2$ is calcium chloride.

5.26   a. Cu$_2$O is copper(I) oxide, or cuprous oxide.
       b. correct
       c. correct
       d. SnI$_2$ is tin(II) iodide, or stannous iodide.
       e. correct
       f. Co$_3$S$_2$ is cobalt(II) sulfide, or cobaltous sulfide.

5.27   CaCO$_3$, calcium carbonate

5.28   Mg(OH)$_2$, magnesium hydroxide

5.29   a. Ca(HCO$_3$)$_2$, calcium hydrogen carbonate   b. NH$_4$Cl, ammonium chloride
       c. Mg(ClO)$_2$, magnesium hypochlorite   d. AgCN, silver cyanide
       e. NH$_4$NO$_3$, ammonium nitrate   f. Al(ClO$_4$)$_3$, aluminum perchlorate
       g. Ba$_3$(PO$_4$)$_2$, barium phosphate   h. KNO$_2$, potassium nitrite
       i. NaC$_2$H$_3$O$_2$, sodium acetate

5.30   a. Na$_2$HPO$_4$, sodium hydrogen phosphate   b. NaOH, sodium hydroxide
       c. Mg(ClO$_2$)$_2$, magnesium chlorite   d. (NH$_4$)$_2$SO$_4$ ammonium sulfate
       e. Ca(MnO$_4$)$_2$, calcium permanganate   f. Mg(ClO$_3$)$_2$, magnesium chlorate
       g. (NH$_4$)$_2$SO$_3$, ammonium sulfite   h. K$_2$CO$_3$, potassium carbonate
       i. Ba(C$_2$H$_3$O$_2$)$_2$, barium acetate

5.31   a. Al(OH)$_3$ b. BaSO$_3$ c. BaSO$_4$
       d. Li$_2$HPO$_4$ e. LiH$_2$PO$_4$ f. Li$_3$PO$_4$

5.32   a. Ca(ClO)$_2$ b. Ca(ClO$_2$)$_2$ c. Ca(ClO$_3$)$_2$
       d. Ca(ClO$_4$)$_2$ e. NH$_4$C$_2$H$_3$O$_2$ f. KMnO$_4$

5.33   nitrogen trihydride

5.34   dihydrogen monoxide (Note: the actual systematic name for water is hydrogen oxide)

5.35   a. dinitrogen monoxide   b. carbon tetrachloride
       c. silicon tetrafluoride   d. nitrogen dioxide
       e. chlorine trifluoride   f. dichlorine heptoxide

5.36   a. arsenic trichloride   b. diarsenic pentoxide
       c. nitrogen triiodide   d. bromine pentafluoride
       e. iodine pentafluoride   f. sulfur trioxide

5.37 a. NBr$_3$ b. XeO$_4$ c. OF$_2$
      d. Cl$_2$O$_7$ e. SF$_6$ f. PI$_3$

5.38 a. N$_2$F$_2$ b. NCl$_3$ c. As$_4$O$_6$
      d. CO   e. CO$_2$ f. N$_2$O$_4$

5.39 hydrogen bromide

5.40 hydrogen iodide

5.41 a. phosphoric acid   b. nitrous acid   c. carbonic acid
      d. nitric acid   e. hypochlorous acid   f. perchloric acid

5.42 a. sulfurous acid   b. sulfuric acid   c. chlorous acid
      d. chloric acid   e. boric acid   f. acetic acid

5.43 a. HNO$_2$ b. HNO$_3$ c. HI
      d. H$_2$SO$_4$ e. H$_2$SO$_3$ f. HBr

5.44 a. HClO   b. H$_3$PO$_4$ c. HClO$_3$
      d. H$_2$S   e. HCl   f. HClO$_4$

■ Solutions to Additional Problems

5.45 NaHCO$_3$, sodium hydrogen carbonate

5.46 Na$_2$CO$_3$, sodium carbonate

5.47 CaCO$_3$, calcium carbonate

5.48 CaO, calcium oxide

5.49 a. FeSO$_4$ b. Fe$_2$(SO$_4$)$_3$
      c. CuC$_2$H$_3$O$_2$ d. Cu(C$_2$H$_3$O$_2$)$_2$

5.50 a. SnF$_2$ b. SnF$_4$ c. Co(NO$_3$)$_2$ d. Co(NO$_3$)$_3$

5.51 a. NiCl$_2$ b. TiF$_4$ c. MnO$_2$ d. Cr$_2$S$_3$

5.52 a. V$_2$O$_5$ b. CrBr$_2$ c. Mn$_2$O$_7$ d. TiCl$_2$

5.53 a. Pb(NO$_3$)$_2$ b. Co(OH)$_2$ c. Cu$_3$(PO$_4$)$_2$
      d. CuNO$_2$ e. Fe(ClO$_2$)$_2$ f. Sn(ClO$_4$)$_4$

5.54 a. $Co(C_2H_3O_2)_3$ b. $Sn(OH)_2$ c. $Cu(NO_3)_2$
         d. $Fe_2(SO_4)_3$ e. $FeSO_4$ f. $FeSO_3$

5.55 a. tin(II) phosphate   b. iron(II) nitrate
         c. chromium(II) sulfate   d. aluminum chlorate

5.56 a. iron(III) hydroxide   b. ammonium phosphate
         c. calcium hypochlorite   d. chromium(III) nitrate

5.57 a. BrO_3^-, bromate ion   b. $N_2O_2^{2-}$, hyponitrite ion
         c. $S_2O_3^{2-}$, thiosulfate ion   d. AsO_4^{3-}, arsenate ion

5.58 a. BrO_2^-, bromite ion   b. BrO_4^-, perbromate ion
         c. IO^-, hypoiodite ion   d. $P_2O_7^{4-}$, diphosphate ion

6. CHEMICAL REACTIONS AND EQUATIONS

■ Solutions to Exercises

6.1 a. Two molecules of hydrogen gas and one molecule of oxygen gas react in the presence of a platinum catalyst to form two molecules of liquid water.

b. Two formula units of aqueous sodium iodide reacts with one molecule of chlorine gas to form two formula units of aqueous sodium chloride and one molecule of aqueous iodine.

c. One molecule of carbon monoxide gas reacts with three molecules of hydrogen gas to form one molecule of methane gas and one molecule of water gas.

d. Two molecules of aqueous hydrogen peroxide react when heated to form two molecules of liquid water and one molecule of oxygen gas.

6.2 a. An analysis shows that the equation is not balanced.

$$NH_4NO_3 \xrightarrow{\Delta} N_2O + H_2O$$

Left side: 2 nitrogen atoms   Right side: 2 nitrogen atoms
           3 oxygen atoms   2 oxygen atoms
           4 hydrogen atoms   2 hydrogen atoms

Begin balancing by placing a temporary "1" in front of NH_4NO_3, the most complicated molecule in the equation.

$$1\ NH_4NO_3 \xrightarrow{\Delta} N_2O + H_2O \qquad \text{(unbalanced)}$$

The nitrogen atoms are already balanced, so put a temporary "1" in front of N_2O.

$$1\ NH_4NO_3 \xrightarrow{\Delta} 1\ N_2O + H_2O \qquad \text{(unbalanced)}$$

CHEMICAL REACTIONS AND EQUATIONS ■ 39

Next, balance the oxygen atoms. Three oxygen atoms are specified on the left side. To obtain three on the right side, you must change the coefficient of H_2O to 2.

$$1\ NH_4NO_3 \xrightarrow{\Delta} 1\ N_2O + 2\ H_2O$$

Finally, balance the hydrogen atoms. There are four hydrogen atoms on the left and four hydrogen atoms on the right. Thus, no change is necessary. As a result, after removing the temporary 1's, you get

$$NH_4NO_3 \xrightarrow{\Delta} N_2O + 2\ H_2O \qquad \text{(balanced)}$$

Checking: Left side: 2 nitrogen atoms   Right side: 2 nitrogen atoms
                      3 oxygen atoms   3 oxygen atoms
                      4 hydrogen atoms   4 hydrogen atoms

The coefficients are whole-numbers and the equation is balanced.

b. The equation is not balanced yet.

$$N_2O_5 \longrightarrow NO_2 + O_2$$

Left side: 2 nitrogen atoms   Right side: 1 nitrogen atom
           5 oxygen atoms   4 oxygen atoms

Begin by placing a temporary "1" in front of N_2O_5, the most complicated molecule.

$$1\ N_2O_5 \longrightarrow NO_2 + O_2 \qquad \text{(unbalanced)}$$

First, balance the nitrogen atoms. There are two nitrogen atoms specified on the left side. To obtain two on the right side, change the coefficient of NO_2 to 2.

$$1\ N_2O_5 \longrightarrow 2\ NO_2 + O_2 \qquad \text{(unbalanced)}$$

Next, balance the oxygen atoms. There are five oxygen atoms specified on the left side and six on the right side. To balance, change the coefficient of O_2 to ½.

$$1\ N_2O_5 \longrightarrow 2\ NO_2 + \tfrac{1}{2}\ O_2$$

The equation is balanced but only whole-number coefficients are permissible. As a result, you must multiply all of the coefficients by two.

$$(2 \times 1)\ N_2O_5 \longrightarrow (2 \times 2)\ NO_2 + (2 \times \tfrac{1}{2})\ O_2$$

Or,

$$2\ N_2O_5 \longrightarrow 4\ NO_2 + O_2 \qquad \text{(balanced)}$$

Checking: Left side: 4 nitrogen atoms   Right side: 4 nitrogen atoms
                      10 oxygen atoms   10 oxygen atoms

c. The equation is not balanced yet.

$$Ca + H_2O \longrightarrow Ca(OH)_2 + H_2 \quad \text{(unbalanced)}$$

Left side: 1 calcium atom
2 hydrogen atoms
1 oxygen atom

Right side: 1 calcium atom
4 hydrogen atoms
2 oxygen atoms

Start by placing a temporary "1" in front of $Ca(OH)_2$, the most complicated substance.

$$Ca + H_2O \longrightarrow 1\ Ca(OH)_2 + H_2 \quad \text{(unbalanced)}$$

First, the calcium atoms are already balanced so place a temporary "1" in front of Ca.

$$1\ Ca + H_2O \longrightarrow 1\ Ca(OH)_2 + H_2 \quad \text{(unbalanced)}$$

Next, balance the oxygen atoms. There are two oxygen atoms specified on the right side and one on the left side. To balance, change the coefficient of H_2O to 2.

$$1\ Ca + 2\ H_2O \longrightarrow 1\ Ca(OH)_2 + H_2$$

Next, balance the hydrogen atoms. There are four hydrogen atoms specified on both sides, so they are already balanced. After removing the temporary 1's, you get

$$Ca + 2\ H_2O \longrightarrow Ca(OH)_2 + H_2 \quad \text{(balanced)}$$

Checking: Left side: 1 calcium atom
4 hydrogen atoms
2 oxygen atoms

Right side: 1 calcium atom
4 hydrogen atoms
2 oxygen atoms

The coefficients are whole-numbers and the equation is balanced.

d. The equation is not balanced yet.

$$As_2S_3 + O_2 \xrightarrow{\Delta} As_2O_3 + SO_2$$

Left side: 2 arsenic atoms
3 sulfur atoms
2 oxygen atoms

Right side: 2 arsenic atoms
1 sulfur atom
5 oxygen atoms

Begin balancing by placing a temporary "1" in front of As_2S_3.

$$1\ As_2S_3 + O_2 \xrightarrow{\Delta} As_2O_3 + SO_2 \quad \text{(unbalanced)}$$

The arsenic atoms are already balanced with two on both sides. Place a temporary "1" in front of As_2O_3.

$$1\ As_2S_3 + O_2 \xrightarrow{\Delta} 1\ As_2O_3 + SO_2 \quad \text{(unbalanced)}$$

Next, balance sulfur atoms. There are three sulfur atoms specified on the left side and one atom on the right. To balance, change the coefficient of SO_2 to 3.

$$1\ As_2S_3\ +\ O_2\ \xrightarrow{\Delta}\ 1\ As_2O_3\ +\ 3\ SO_2 \qquad \text{(unbalanced)}$$

Next, balance oxygen atoms. There are nine oxygen atoms specified on the right side and two on the left side. To balance, change the coefficient of O_2 to 9/2.

$$1\ As_2S_3\ +\ 9/2\ O_2\ \xrightarrow{\Delta}\ 1\ As_2O_3\ +\ 3\ SO_2$$

The equation is balanced but only whole-number coefficients are permissible. As a result, you must multiply all of the coefficients by two.

$$(2\times 1)\ As_2S_3\ +\ (2\times 9/2)\ O_2\ \xrightarrow{\Delta}\ (2\times 1)\ As_2O_3\ +\ (2\times 3)\ SO_2$$

Or,

$$2\ As_2S_3\ +\ 9\ O_2\ \xrightarrow{\Delta}\ 2\ As_2O_3\ +\ 6\ SO_2 \qquad \text{(balanced)}$$

Checking: Left side: 4 arsenic atoms   Right side: 4 arsenic atoms
                      6 sulfur atoms   6 sulfur atoms
                      18 oxygen atoms   18 oxygen atoms

The coefficients are whole-numbers and the equation is balanced.

6.3 Begin by writing an unbalanced equation that agrees with the description. Analyze the numbers of atoms of each element.

$$Fe\ +\ O_2\ \longrightarrow\ Fe_2O_3 \qquad \text{(unbalanced)}$$

Left side: 1 iron atom   Right side: 2 iron atoms
            2 oxygen atoms   3 oxygen atoms

Fe_2O_3 is the most complicated molecule. Careful analysis shows that if you put a one in front of it, you would need a 3/2 in front of O_2, which you would need to eliminate. Anticipate this situation by placing a 2 in front of Fe_2O_3.

$$Fe\ +\ O_2\ \longrightarrow\ 2\ Fe_2O_3 \qquad \text{(unbalanced)}$$

Next, balance the iron atoms. There are four iron atoms on the right side. To balance, put a 4 in front of Fe.

$$4\ Fe\ +\ O_2\ \longrightarrow\ 2\ Fe_2O_3 \qquad \text{(unbalanced)}$$

Finally, balance oxygen atoms. There are six oxygen atoms on the right side. To balance, put a 3 in front of O_2.

$$4\ Fe + 3\ O_2 \longrightarrow 2\ Fe_2O_3 \quad (balanced)$$

Checking: Left side: 4 iron atoms   Right side: 4 iron atoms
6 oxygen atoms   6 oxygen atoms

The equation is balanced.

6.4 a. This is a decomposition reaction since a compound has decomposed into an element and a compound.

  b. This is a combination reaction since two elements combine to form a third substance.

  c. Since this reaction is not represented by either A + B ⟶ AB or AB ⟶ A + B, it is neither a combination reaction nor a decomposition reaction.

6.5 The equation: $2\ P(s) + 5\ Cl_2(g) \longrightarrow 2\ PCl_5(s)$
does not represent a single replacement reaction because it cannot be represented by A + BC ⟶ AB + C. It is a combination reaction because it can be represented by A + B ⟶ AB.

6.6 Since, according to Table 6.2, carbonates are insoluble, calcium carbonate will form a precipitate. The balanced equation for the chemical reaction is

$$(NH_4)_2CO_3(aq) + CaBr_2(aq) \longrightarrow CaCO_3(s) + 2\ NH_4Br(aq)$$

Also, according to Table 6.2, ammonium compounds are soluble, so NH_4^+ and Br^- are the spectator ions.

6.7 Limestone ($CaCO_3$) is insoluble so "(s)" must be added after it, and nitric acid (HNO_3) is an aqueous solution so "(aq)" is required after it. The balanced chemical equation is

$$CaCO_3(s) + 2\ HNO_3(aq) \longrightarrow Ca(NO_3)_2(aq) + CO_2(g) + H_2O(l)$$

For the products, $Ca(NO_3)_2$ is (aq), since nitrates are soluble according to Table 6.2. The gas that forms is CO_2, according to Table 6.3. The final product is water, which is a liquid.

CHEMICAL REACTIONS AND EQUATIONS ■ 43

6.8 Start by writing as much of the equation as you can. Calcium hydroxide is soluble, according to Table 6.2, and sulfuric acid is an aqueous solution. The product, calcium sulfate ($CaSO_4$) is also insoluble according to Table 6.2, so it is the precipitate. The final product is water. This gives

$$Ca(OH)_2(aq) + H_2SO_4(aq) \longrightarrow CaSO_4(s) + 2\ H_2O(l)$$

6.9 a. The balanced chemical equation is

$$C_3H_8(g) + 5\ O_2(g) \longrightarrow 3\ CO_2(g) + 4\ H_2O(g)$$

This is a combustion reaction because of the reaction with oxygen (O_2) and the formation of CO_2 and H_2O as products.

b. The balanced equation is

$$Mg(s) + 2\ HBr(aq) \longrightarrow MgBr_2(aq) + H_2(g)$$

This is a single replacement reaction because the element magnesium (Mg) is replacing the element hydrogen (H) in the compound HBr.

c. The balanced equation is

$$CS_2(l) + 3\ O_2(g) \longrightarrow CO_2(g) + 2\ SO_2(g)$$

This may be a combustion reaction because of the reaction with oxygen (O_2).

■ Answers to Questions to Test Your Reading

6.1 A <u>chemical reaction</u> is a change in which one or more kinds of matter are transformed into one or more new kinds of matter. The <u>reactants</u> are the chemicals that are present before a chemical reaction, and the <u>products</u> are the chemicals that result from the chemical reaction.

6.2 A <u>chemical equation</u> is a symbolic way of expressing a chemical reaction. A <u>coefficient</u> is a number in front of a formula in the chemical equation. They tell you how many molecules of each reactant are involved and how many molecules of each product are formed. An example of a balanced equation is

$$CH_4 + 2\ O_2 \longrightarrow CO_2 + 2\ H_2O$$

One molecule of methane and two molecules of oxygen react to form one molecule of carbon dioxide and two molecules of water.

6.3 A <u>combination reaction</u> is a reaction in which two substances combine to form a third. An example is

$$4\ Fe(s) + 3\ O_2(g) \longrightarrow 2\ Fe_2O_3(s)$$

A combination reaction is the opposite of a decomposition reaction.

6.4 A <u>decomposition reaction</u> is a reaction in which a single compound decomposes to form two or more other substances. The products of this type of reaction may be two or more elements, but it is not always this way. For example, the decomposition of mercury (II) oxide forms two elements.

$$2\ HgO(s) \xrightarrow{\Delta} 2\ Hg(l) + O_2(g)$$

However, the decomposition of limestone produces two compounds.

$$CaCO_3(s) \xrightarrow{\Delta} CaO(s) + CO_2(g)$$

6.5 A <u>single replacement reaction</u> is a reaction in which an element reacts with a compound and replaces another element in the compound. An example is

$$Zn(s) + 2\ HCl(aq) \longrightarrow ZnCl_2(aq) + H_2(g)$$

6.6 A <u>double replacement reaction</u> is a reaction in which two compounds exchange parts so that two new compounds are formed. They are the same as <u>metathesis reactions</u>. There is no difference.

6.7 The three different ways that double replacement reaction is driven in the direction of the arrow are, first, a solid precipitate forms. An example is

$$2\ KI(aq) + Pb(NO_3)_2(aq) \longrightarrow 2\ KNO_3(aq) + PbI_2(s)$$

Second, a gas forms. An example is

$$HC_2H_3O_2(aq) + NaHCO_3(aq) \longrightarrow NaC_2H_3O_2(aq) + CO_2(g) + H_2O(l)$$

Third, water forms in a neutralization reaction. An example is

$$HCl(aq) + NaOH(aq) \longrightarrow H_2O(l) + NaCl(aq)$$

6.8 An <u>acid</u> is a compound that produces hydrogen ions (H$^+$) when it is dissolved in water. A <u>base</u> is a compound that produces hydroxide ions (OH$^-$) when it is dissolved in water. A <u>neutralization reaction</u> is the reaction that occurs between an acid and a base with the formation of an ionic compound and usually water. An example is

$$HCl(aq) + NaOH(aq) \longrightarrow H_2O(l) + NaCl(aq)$$

6.9 <u>Spectator ions</u> are ions that do not take part in an ionic reaction. In the following reaction

$$2\ KI(aq) + Pb(NO_3)_2(aq) \longrightarrow 2\ KNO_3(aq) + PbI_2(s)$$

since potassium nitrate (KNO$_3$) is soluble, the ions K$^+$ and NO$_3^-$ do not take part in the reaction and are the spectator ions.

6.10 A <u>combustion reaction</u> is a reaction of a substance with either pure oxygen or oxygen in the air with the with the rapid release of heat and the appearance of a flame. Not all combustion reactions are combination reactions, but some are. An example of a combination reaction that is also a combustion reaction is

$$C(s) + O_2(g) \longrightarrow CO_2(g)$$

An example that is not a combination reaction is

$$CH_4(g) + 2\ O_2(g) \longrightarrow CO_2(g) + 2\ H_2O(g)$$

■ Solutions to Practice Problems

6.11 a. Two atoms of solid potassium and one molecule of liquid bromine react to form two formula units of solid potassium bromide.

b. One molecule of solid phosphorus and six molecules of chlorine gas react to form four molecules of liquid phosphorus trichloride.

6.12 a. One atom of solid calcium and two molecules of liquid water react to form one formula unit of aqueous calcium hydroxide and one molecule of hydrogen gas.

b. One molecule of liquid ethanol and three molecules of oxygen gas react to form two molecules of carbon dioxide gas and three molecules of water gas.

6.13 a. One formula unit of solid barium carbonate reacts when heated to produce one formula unit of solid barium oxide and one molecule of carbon dioxide gas.

  b. Two molecules of aqueous hydrogen peroxide react in the presence of a potassium iodide catalyst to produce two molecules of liquid water and one molecule of oxygen gas.

6.14 a. Two formula units of aqueous potassium nitrate react when heated to produce two formula units of aqueous potassium nitrite and one molecule of oxygen gas.

  b. Two molecules of carbon monoxide gas react in the presence of a platinum catalyst to form two molecules of carbon dioxide gas.

6.15 a. $2\ CH_3OH + 3\ O_2 \longrightarrow 2\ CO_2 + 4\ H_2O$

  b. $2\ Mg + SiO_2 \longrightarrow 2\ MgO + Si$

  c. $Cl_2O_7 + H_2O \longrightarrow 2\ HClO_4$

  d. $TiCl_4 + 2\ H_2O \longrightarrow TiO_2 + 4\ HCl$

6.16 a. $C_2H_5OH + 3\ O_2 \longrightarrow 2\ CO_2 + 3\ H_2O$

  b. $Fe_2O_3 + 3\ H_2 \longrightarrow 2\ Fe + 3\ H_2O$

  c. $Al_2S_3 + 6\ H_2O \longrightarrow 2\ Al(OH)_3 + 3\ H_2S$

  d. $C_5H_{12} + 8\ O_2 \longrightarrow 5\ CO_2 + 6\ H_2O$

6.17 a. $CO + 3\ H_2 \longrightarrow CH_4 + H_2O$

  b. $Ba + 2\ H_2O \longrightarrow Ba(OH)_2 + H_2$

c. $2 H_2S + 3 O_2 \longrightarrow 2 H_2O + 2 SO_2$

d. $4 H_3PO_3 \longrightarrow 3 H_3PO_4 + PH_3$

6.18 a. $V_2O_5 + 2 H_2 \longrightarrow V_2O_3 + 2 H_2O$

b. $2 C_6H_6 + 15 O_2 \longrightarrow 12 CO_2 + 6 H_2O$

c. $Al_4C_3 + 12 H_2O \longrightarrow 4 Al(OH)_3 + 3 CH_4$

d. $4 NH_3 + 5 O_2 \longrightarrow 4 NO + 6 H_2O$

6.19 $N_2(g) + O_2(g) \longrightarrow 2 NO(g)$

6.20 $2 NO(g) + O_2(g) \longrightarrow 2 NO_2(g)$

6.21 $4 NO_2(g) + O_2(g) \longrightarrow 2 N_2O_5(g)$

6.22 $2 N_2O_5(g) \xrightarrow[Pt]{\Delta} 2 N_2(g) + 5 O_2(g)$

6.23 a. Neither a combination nor a decomposition reaction.

b. Combination reaction.

c. Decomposition reaction.

d. Combination reaction.

6.24 a. Decomposition reaction.

b. Combination reaction.

c. Combination reaction.

d. Decomposition reaction.

6.25 a. $2\ CH_4O(l) + 3\ O_2(g) \longrightarrow 2\ CO_2(g) + 4\ H_2O(g)$

Neither a decomposition reaction nor a combination reaction.

b. $2\ NO_2(g) \longrightarrow 2\ NO(g) + O_2(g)$

Decomposition reaction.

c. $N_2(g) + 3\ H_2(g) \longrightarrow 2\ NH_3(g)$

Combination reaction.

d. $4\ Li(s) + O_2(g) \longrightarrow 2\ Li_2O(g)$

Combination reaction.

6.26 a. $C_2H_6O(l) + 2\ O_2(g) \longrightarrow 2\ CO(g) + 3\ H_2O(g)$

Neither a decomposition reaction nor a combination reaction.

b. $C_2H_6O(l) + 3\ O_2(g) \longrightarrow 2\ CO_2(g) + 3\ H_2O(g)$

Neither a decomposition reaction nor a combination reaction.

c. $P_4(s) + 6\ Cl_2(g) \longrightarrow 4\ PCl_3(l)$

Combination reaction.

d. $N_2(g) + 3\ I_2(g) \longrightarrow 2\ NI_3(g)$

Combination reaction.

6.27 a. $Cd(s) + 2\ AgNO_3(aq) \longrightarrow Cd(NO_3)_2(aq) + 2\ Ag(s)$

Single replacement reaction.

b. $C_3H_6(g) + H_2(g) \longrightarrow C_3H_8(g)$

Not a single replacement reaction.

6.28 a. $O_2(g) + 2 F_2(g) \longrightarrow 2 OF_2(g)$

Not a single replacement reaction.

b. $3 Mg(s) + Fe_2(SO_4)_3(aq) \longrightarrow 3 MgSO_4(aq) + 2 Fe(s)$

Single replacement reaction.

6.29 Exchange cations and anions so that the products can be determined. In this case, the products would be sodium nitrate and lead carbonate. According to Table 6.2, lead carbonate is insoluble, and will form a precipitate. Also, sodium nitrate is soluble, according to the table. The reaction is

$$Pb(NO_3)_2(aq) + Na_2CO_3(aq) \longrightarrow PbCO_3(s) + 2 NaNO_3(aq)$$

Since sodium nitrate is soluble, Na^+ and NO_3^- are the spectator ions.

6.30 Exchange cations and anions so that the products can be determined. In this case, the products would be nickel (II) sulfide and sodium chloride. According to Table 6.2, nickel (II) sulfide is insoluble and will form a precipitate. Also, sodium chloride is soluble, according to the table. The reaction is

$$NiCl_2(aq) + Na_2S(aq) \longrightarrow NiS(s) + 2 NaCl(aq)$$

Since sodium chloride is soluble, Na^+ and Cl^- are the spectator ions.

6.31 a. $MgSO_4(aq) + 2 NaOH(aq) \longrightarrow Mg(OH)_2(s) + Na_2SO_4(aq)$

The precipitate is $Mg(OH)_2$ and the spectator ions are Na^+ and SO_4^{2-}.

b. No precipitate will form.

c. No precipitate will form.

d. No precipitate will form.

6.32 a. $Ba(NO_3)_2(aq) + K_2SO_4(aq) \longrightarrow BaSO_4(s) + 2 KNO_3(aq)$

The precipitate is $BaSO_4$ and the spectator ions are K^+ and NO_3^-.

b. No precipitate will form.

c. $AgNO_3(aq) + NaI(aq) \longrightarrow AgI(s) + NaNO_3(aq)$

The precipitate is AgI and the spectator ions are Na^+ and NO_3^-.

d. $Pb(NO_3)_2(aq) + 2\ NH_4Cl(aq) \longrightarrow PbCl_2(s) + 2\ NH_4NO_3(aq)$

The precipitate is $PbCl_2$ and the spectator ions are NH_4^+ and NO_3^-.

6.33 $Na_2CO_3(aq) + 2\ HC_2H_3O_2(aq) \longrightarrow 2\ NaC_2H_3O_2(aq) + CO_2(g) + H_2O(l)$

The gas that forms is CO_2 and the spectator ions are Na^+ and $C_2H_3O_2^-$.

6.34 $Li_2SO_3(aq) + 2\ HBr(aq) \longrightarrow 2\ LiBr(aq) + SO_2(g) + H_2O(l)$

The gas that forms is SO_2 and the spectator ions are Li^+ and Br^-.

6.35 a. No gas will form.

b. No gas will form.

c. $(NH_4)_2CO_3(aq) + 2\ HCl(aq) \longrightarrow 2\ NH_4Cl(aq) + CO_2(g) + H_2O(l)$

The gas that forms is CO_2 and the spectator ions are NH_4^+ and Cl^-.

d. No gas will form.

6.36 a. $MnS(s) + H_2SO_4(aq) \longrightarrow MnSO_4(aq) + H_2S(g)$

The gas that forms is H_2S and the spectator ion is SO_4^{2-}.

b. No gas will form.

c. $CaCO_3(s) + 2\ HNO_3(aq) \longrightarrow Ca(NO_3)_2(aq) + CO_2(g) + H_2O(l)$

The gas that forms is CO_2 and the spectator ion is NO_3^-.

d. No gas will form.

6.37 a. base   b. acid   c. base   d. acid   e. neither
f. neither   g. neither   h. base   i. neither

6.38 a. acid   b. base   c. neither   d. acid   e. neither
        f. base   g. neither   h. acid   i. acid

6.39 a. HNO$_3$(aq) + NaOH(aq) $\longrightarrow$ NaNO$_3$(aq) + H$_2$O(l)

b. Ba(OH)$_2$(aq) + 2 HCl(aq) $\longrightarrow$ BaCl$_2$(aq) + 2 H$_2$O(l)

c. LiOH(aq) + HBr(aq) $\longrightarrow$ LiBr(aq) + H$_2$O(l)

d. HC$_2$H$_3$O$_2$(aq) + NaOH(aq) $\longrightarrow$ NaC$_2$H$_3$O$_2$(aq) + H$_2$O(l)

6.40 a. Al(OH)$_3$(s) + 3 HNO$_3$(aq) $\longrightarrow$ Al(NO$_3$)$_3$(aq) + 3 H$_2$O(l)

b. H$_2$SO$_4$(aq) + 2 KOH(aq) $\longrightarrow$ K$_2$SO$_4$(aq) + 2 H$_2$O(l)

c. NaOH(aq) + HI(aq) $\longrightarrow$ NaI(aq) + H$_2$O(l)

d. 2 HClO$_4$(aq) + Ba(OH)$_2$(aq) $\longrightarrow$ Ba(ClO$_4$)$_2$(aq) + 2 H$_2$O(l)

6.41 The reaction of octane (C$_8$H$_{18}$) with oxygen is a combustion reaction. The balanced equation is

$$2\ C_8H_{18}(l) + 25\ O_2(g) \longrightarrow 16\ CO_2(g) + 18\ H_2O(g)$$

6.42 The reaction of butane (C$_4$H$_{10}$) with oxygen is a combustion reaction. The balanced equation is

$$2\ C_4H_{10}(g) + 13\ O_2(g) \longrightarrow 8\ CO_2(g) + 10\ H_2O(g)$$

Solutions to Additional Problems

6.43 $NH_4Cl(s) \xrightarrow{\Delta} NH_3(g) + HCl(g)$

6.44 $2\ Na(s) + 2\ H_2O(l) \longrightarrow 2\ NaOH(aq) + H_2(g)$

6.45 $6\ CO_2 + 6\ H_2O \longrightarrow C_6H_{12}O_6 + 6\ O_2$

6.46 $C_{12}H_{22}O_{11} + H_2O \longrightarrow 4\ C_2H_6O + 4\ CO_2$

6.47 $2\ NaC_{18}H_{36}O_2(aq) + CaCl_2(aq) \longrightarrow Ca(C_{18}H_{36}O_2)_2(s) + 2\ NaCl(aq)$

6.48 $C_{12}H_{22}O_{11}(s) \xrightarrow{\Delta} 12\ C(s) + 11\ H_2O(g)$

6.49 $6\ Li(s) + N_2(g) \longrightarrow 2\ Li_3N(s)$

6.50 $3\ Mg(s) + N_2(g) \longrightarrow Mg_3N_2(s)$

6.51 a. $HNO_3(aq) + NaOH(aq) \longrightarrow NaNO_3(aq) + H_2O(l)$

The reaction is driven by the formation of water.

b. $HNO_3(aq) + NaCN(aq) \longrightarrow NaNO_3(aq) + HCN(g)$

The reaction is driven by the formation of a gas.

c. $Pb(NO_3)_2(aq) + 2\ NaCl(aq) \longrightarrow 2\ NaNO_3(aq) + PbCl_2(s)$

The reaction is driven by the formation of a precipitate.

6.52 a. $Ni(NO_3)_2(aq) + 2\ NaOH(aq) \longrightarrow 2\ NaNO_3(aq) + Ni(OH)_2(s)$

The reaction is driven by the formation of a precipitate.

CHEMICAL REACTIONS AND EQUATIONS   53

b. $2\ HCl(aq) + Na_2SO_3(aq) \longrightarrow 2\ NaCl(aq) + H_2O(l) + SO_2(g)$

The reaction is driven by the formation of a gas.

c. $H_3PO_4(aq) + 3\ KOH(aq) \longrightarrow K_3PO_4(aq) + 3\ H_2O(l)$

The reaction is driven by the formation of water.

6.53 a. soluble   b. soluble   c. insoluble   d. insoluble   e. insoluble
f. soluble   g. insoluble   h. soluble   i. insoluble

6.54 a. insoluble   b. insoluble   c. insoluble   d. soluble   e. soluble
f. insoluble   g. soluble   h. insoluble   i. soluble

6.55 a. Single replacement reaction:

$$2\ Na(s) + 2\ H_2O(l) \longrightarrow 2\ NaOH(aq) + H_2(g)$$

b. Double replacement reaction:

$$CrCl_3(aq) + 3\ NaOH(aq) \longrightarrow Cr(OH)_3(s) + 3\ NaCl(aq)$$

c. Single replacement reaction:

$$Zn(s) + CuSO_4(aq) \longrightarrow ZnSO_4(aq) + Cu(s)$$

d. Combustion reaction:

$$C_6H_{12}(l) + 9\ O_2(g) \longrightarrow 6\ CO_2(g) + 6\ H_2O(g)$$

6.56 a. Double replacement reaction (neutralization reaction):

$$3\ HI(aq) + Al(OH)_3(s) \longrightarrow 3\ H_2O(l) + AlI_3(aq)$$

b. Single replacement reaction:

$$2\ Al(s) + 2\ H_3PO_4(aq) \longrightarrow 2\ AlPO_4(s) + 3\ H_2(g)$$

c. Decomposition reaction:

$$CaSO_3(s) \longrightarrow CaO(s) + SO_2(g)$$

d. Combination reaction:

$$4\ Al(s) + 3\ O_2(g) \longrightarrow 2\ Al_2O_3(s)$$

6.57 a. $BaCO_3(s) + 2\ HNO_3(aq) \longrightarrow Ba(NO_3)_2(aq) + CO_2(g) + H_2O(l)$

b. $BaCl_2(aq) + H_2SO_4(aq) \longrightarrow BaSO_4(s) + 2\ HCl(aq)$

c. $Ca(OH)_2(s) + 2\ HC_2H_3O_2(aq) \longrightarrow Ca(C_2H_3O_2)_2(aq) + 2\ H_2O(l)$

d. $2\ Al(s) + 3\ NiSO_4(aq) \longrightarrow Al_2(SO_4)_3(aq) + 3\ Ni(s)$

6.58 a. $Ni(OH)_2(s) + 2\ HCl(aq) \longrightarrow NiCl_2(aq) + 2\ H_2O(l)$

b. $MnCl_2(aq) + 2\ NaOH(aq) \longrightarrow Mn(OH)_2(s) + 2\ NaCl(aq)$

c. $3\ KOH(aq) + H_3PO_4(aq) \longrightarrow K_3PO_4(aq) + 3\ H_2O(l)$

d. $MnS_2(s) + 2\ H_2SO_4(aq) \longrightarrow Mn(SO_4)_2(aq) + 2\ H_2S(g)$

6.59 a. $2\ Li(s) + Cl_2(g) \longrightarrow 2\ LiCl(s)$

b. $2\ C_{10}H_{22}(s) + 31\ O_2(g) \longrightarrow 20\ CO_2(g) + 22\ H_2O(g)$

c. The combustion of decane ($C_{10}H_{22}$), in the presence of a deficiency of oxygen may form the following products: carbon (soot), carbon monoxide (CO), carbon dioxide (CO_2), and water (H_2O). The proportions of each product formed cannot be specified.

d. $2\ Au_2O_3(s) \xrightarrow{\Delta} 4\ Au(s) + 3\ O_2(g)$

6.60 a. $H_2SO_3(aq) \longrightarrow SO_2(g) + H_2O(l)$

b. $NaOH(aq) + HClO_4(aq) \longrightarrow NaClO_4(aq) + H_2O(l)$

c. $Na_2O(s) + SO_3(g) \longrightarrow Na_2SO_4(s)$

d. $C_6H_6O(l) + 7\ O_2(g) \longrightarrow 6\ CO_2(g) + 3\ H_2O(g)$

6.61 $2\ PbS(s) + 3\ O_2(g) \xrightarrow{\Delta} 2\ PbO(s) + 2\ SO_2(g)$

6.62 $4\ NH_3(g) + 5\ O_2(g) \xrightarrow{Pt} 4\ NO(g) + 6\ H_2O(g)$

6.63 $Fe_2O_3(s) + 3\ CO(g) \longrightarrow 2\ Fe(l) + 3\ CO_2(g)$

6.64 $Pb(s) + PbO_2(s) + 2\ H_2SO_4(aq) \longrightarrow 2\ PbSO_4(aq) + 2\ H_2O(l)$

7. CHEMICAL COMPOSITION

■ Solutions to Exercises

<u>Note on significant figures:</u> The first time the final answer is written, it is written first with one nonsignificant figure. The least significant digit is also underlined. The final answer is then rounded to the correct number of significant figures but the least significant digit is no longer underlined. Also, in these problems, AW stands for atomic weight.

7.1 The molecular weight of $C_2H_2F_4$ is the sum of the atomic weights of all of the atoms in the formula of the compound. This gives

2 x AW for carbon	= 2 x	12.01 amu =	24.02 amu
2 x AW for hydrogen	= 2 x	1.008 amu =	2.016 amu
4 x AW for fluorine	= 4 x	19.00 amu =	76.00 amu

Molecular weight of $C_2H_2F_4$ = 102.0$\underline{3}$6 amu = 102.04 amu

7.2 $Mg(OH)_2$ has the formula weight

1 x AW for magnesium	= 1 x	24.31 amu =	24.31 amu
2 x AW for oxygen	= 2 x	16.00 amu =	32.00 amu
2 x AW for hydrogen	= 2 x	1.008 amu =	2.016 amu

Molecular weight of $Mg(OH)_2$ = 58.3$\underline{2}$6 amu = 58.33 amu

PCl_5 has the formula weight

1 x AW for phosphorus	= 1 x	30.97 amu =	30.97 amu
5 x AW for chlorine	= 5 x	35.45 amu =	177.25 amu

Molecular weight of PCl_5 = 208.2$\underline{2}$ amu = 208.22 amu

CHEMICAL COMPOSITION ■ 57

7.3 First, convert moles of glucose to molecules using the conversion factor

$$\frac{6.022 \times 10^{23} \text{ molecules } C_6H_{12}O_6}{1 \text{ mol } C_6H_{12}O_6}$$

This gives

$$0.225 \text{ mol } C_6H_{12}O_6 \times \frac{6.022 \times 10^{23} \, C_6H_{12}O_6 \text{ molecules}}{1 \text{ mol } C_6H_{12}O_6} = 1.355 \times 10^{23}$$

$$= 1.36 \times 10^{23} \, C_6H_{12}O_6 \text{ molecules}$$

In terms of the individual elements, there are 6 carbon atoms, 12 hydrogen atoms, and 6 oxygen atoms in each glucose molecule. Using these numbers as conversion factors gives

$$1.36 \times 10^{23} \, C_6H_{12}O_6 \text{ molecules} \times \frac{6 \text{ C atoms}}{1 \, C_6H_{12}O_6 \text{ molecule}} = 8.16 \times 10^{23} \text{ C atoms}$$

$$1.36 \times 10^{23} \, C_6H_{12}O_6 \text{ molecules} \times \frac{12 \text{ H atoms}}{1 \, C_6H_{12}O_6 \text{ molecule}} = 1.63 \times 10^{24} \text{ H atoms}$$

$$1.36 \times 10^{23} \, C_6H_{12}O_6 \text{ molecules} \times \frac{6 \text{ O atoms}}{1 \, C_6H_{12}O_6 \text{ molecule}} = 8.16 \times 10^{23} \text{ O atoms}$$

7.4 Iodine occurs as a two-atom molecule. The molar mass of iodine is double its atomic weight expressed in grams, or

$$2 \times 126.90 \text{ g} = 253.80 \text{ g}$$

Radium is a Group IIA element. Its molar mass is its atomic weight expressed in grams, or 226.03 g.

7.5 The molar mass of sodium phosphate, Na_3PO_4, is its formula weight expressed in grams, or

3 x AW for sodium	= 3 x	22.99 amu =	68.97 amu
1 x AW for phosphorus	= 1 x	30.97 amu =	30.97 amu
4 x AW for oxygen	= 4 x	16.00 amu =	64.00 amu
Formula weight of Na_3PO_4		=	163.94 amu

Thus, the molar mass of Na_3PO_4 is 163.94 g.

58 ■ CHAPTER 7

7.6 Magnesium iodide, MgI_2, is an ionic compound. Its formula weight is

$$
\begin{aligned}
1 \times \text{AW for magnesium} &= 1 \times 24.31 \text{ amu} = 24.31 \text{ amu} \\
2 \times \text{AW for iodine} &= 2 \times 126.90 \text{ amu} = 253.80 \text{ amu} \\
\hline
\text{Formula weight of } MgI_2 &= 278.11 \text{ amu}
\end{aligned}
$$

Thus, the molar mass of MgI_2. The amount of magnesium iodide present in the sample is

$$53.8 \text{ g } MgI_2 \times \frac{1 \text{ mol } MgI_2}{278.11 \text{ g } MgI_2} = 0.1934 = 0.193 \text{ mol } MgI_2$$

7.7 Water, H_2O, is a molecule. Its molecular weight is

$$
\begin{aligned}
2 \times \text{AW for hydrogen} &= 2 \times 1.008 \text{ amu} = 2.016 \text{ amu} \\
1 \times \text{AW for oxygen} &= 1 \times 16.00 \text{ amu} = 16.00 \text{ amu} \\
\hline
\text{Molecular weight of } H_2O &= 18.016 \text{ amu}
\end{aligned}
$$

Thus, the molar mass of water is 18.016 g. The mass of water in the drop is

$$0.00278 \text{ mol } H_2O \times \frac{18.016 \text{ g } H_2O}{1 \text{ mol } H_2O} = 0.05008 = 0.0501 \text{ g } H_2O$$

7.8 The molar mass of water is 18.016 g (see Exercise 7.7 for the calculation). This quantity of water contains 2.016 g of hydrogen and 16.00 g of oxygen. The mass percentages are thus

$$\text{mass percent of hydrogen} = \frac{2.016 \text{ g}}{18.016 \text{ g}} \times 100\% = 11.19\% = 11.2\%$$

$$\text{mass percent of oxygen} = \frac{16.00 \text{ g}}{18.016 \text{ g}} \times 100\% = 88.81\% = 88.8\%$$

7.9 Step 1. Calculate the molar mass of each compound. For NH_4NO_3, the formula weight is

$$
\begin{aligned}
2 \times \text{AW for nitrogen} &= 2 \times 14.01 \text{ amu} = 28.02 \text{ amu} \\
4 \times \text{AW for hydrogen} &= 4 \times 1.008 \text{ amu} = 4.032 \text{ amu} \\
3 \times \text{AW for oxygen} &= 3 \times 16.00 \text{ amu} = 48.00 \text{ amu} \\
\hline
\text{Formula weight of } NH_4NO_3 &= 80.052 \text{ amu}
\end{aligned}
$$

Thus, the molar mass of NH_4NO_3 is 80.052 g.

CHEMICAL COMPOSITION ■ 59

For NH$_2$CONH$_2$, the formula weight is

2 x AW for nitrogen	=	2 x 14.01 amu =	28.02 amu
4 x AW for hydrogen	=	4 x 1.008 amu =	4.032 amu
1 x AW for carbon	=	1 x 12.01 amu =	12.01 amu
1 x AW for oxygen	=	1 x 16.00 amu =	16.00 amu

$$\text{Formula weight of NH}_2\text{CONH}_2 = 60.062 \text{ amu}$$

Thus, the molar mass of NH$_2$CONH$_2$ is 60.062 g.

Step 2. Decide how many grams of nitrogen are in each molar mass. In this case, both compounds contain 28.02 g of nitrogen per mole.

Step 3. Calculate the mass percentage of nitrogen in each compound.

$$\text{mass percent nitrogen in NH}_4\text{NO}_3 = \frac{28.02 \text{ g}}{80.052 \text{ g}} \times 100\% = 35.002\% = 35.00\%$$

$$\text{mass percent nitrogen in NH}_2\text{CONH}_2 = \frac{28.02 \text{ g}}{60.062 \text{ g}} \times 100\% = 46.652\% = 46.65\%$$

Therefore, NH$_2$CONH$_2$ contains more nitrogen per gram.

7.10 Step 1. In 100 g of methanol there is 37.5 g of carbon, 12.6 g of hydrogen, and 49.9 g of oxygen.

Step 2. Convert these quantities to moles using the molar masses.

$$\text{moles of carbon} = 37.5 \text{ g C} \times \frac{1 \text{ mol C}}{12.01 \text{ g C}} = 3.122 \text{ mol C}$$

$$\text{moles of hydrogen} = 12.6 \text{ g H} \times \frac{1 \text{ mol H}}{1.008 \text{ g H}} = 12.50 \text{ mol H}$$

$$\text{moles of oxygen} = 49.9 \text{ g O} \times \frac{1 \text{ mol O}}{16.00 \text{ g O}} = 3.119 \text{ mol O}$$

60 ■ CHAPTER 7

Step 3. Divide each quantity by the smallest number of moles to obtain the subscripts.

$$\text{subscript for carbon} = \frac{3.122 \text{ mol}}{3.119 \text{ mol}} = 1.00$$

$$\text{subscript for hydrogen} = \frac{12.50 \text{ mol}}{3.119 \text{ mol}} = 4.01$$

$$\text{subscript for oxygen} = \frac{3.119 \text{ mol}}{3.119 \text{ mol}} = 1.00$$

Since each subscript is a whole number, the empirical formula of methanol is CH_4O.

7.11 First, calculate the empirical formula weight of C_3H_6O.

3 x AW for carbon	= 3 x	12.01 amu =	36.03 amu
6 x AW for hydrogen	= 6 x	1.008 amu =	6.048 amu
1 x AW for oxygen	= 1 x	16.00 amu =	16.00 amu

$$\text{Empirical formula weight of } C_3H_6O = 58.078 \text{ amu}$$

Next, calculate n.

$$n = \frac{\text{molecular weight}}{\text{empirical formula weight}} = \frac{116 \text{ amu}}{58.078 \text{ amu}} = 2.00$$

Thus, the molecular formula of ethyl acetate is $(C_3H_6O)_2$, or $C_6H_{12}O_2$.

7.12 First, find the empirical formula of the compound.

Step 1. 100 g of the compound contains 92.2 g carbon and 7.8 g of hydrogen.

Step 2. Convert to moles using the molar masses as conversion factors.

$$\text{mol C} = 92.2 \text{ g C} \times \frac{1 \text{ mol C}}{12.01 \text{ g C}} = 7.677 \text{ mol C}$$

$$\text{mol H} = 7.8 \text{ g H} \times \frac{1 \text{ mol H}}{1.008 \text{ g H}} = 7.74 \text{ mol H}$$

Step 3. Divide each quantity by the smallest number of moles to obtain the subscripts.

$$\text{subscript for carbon} = \frac{7.677 \text{ mol}}{7.677 \text{ mol}} = 1.00$$

$$\text{subscript for hydrogen} = \frac{7.74 \text{ mol}}{7.677 \text{ mol}} = 1.01$$

Since the subscripts are both whole numbers, the empirical formula is CH, with an empirical formula weight of

1 x AW for carbon	=	1 x	12.01 amu =	12.01 amu	
1 x AW for hydrogen	=	1 x	1.008 amu =	1.008 amu	
		Empirical formula weight of CH	=	13.018 amu	

Since the molecular weight is 39.1 amu, you can calculate n.

$$n = \frac{\text{molecular weight}}{\text{empirical formula weight}} = \frac{39.1 \text{ amu}}{13.018 \text{ amu}} = 3.00$$

Thus, the molecular formula of the compound is $(CH)_3$, or C_3H_3.

■ Answers to Questions to Test Your Reading

7.1 The <u>molecular weight</u> of a substance is the sum of the atomic weights of all of the atoms in the formula of the molecule. This term is used for molecular compounds. <u>Formula weight</u> is the sum of the atomic weights of all of the atoms in a formula unit of a substance. It is used mostly with ionic compounds.

7.2 Molecular compounds can be expressed as either a molecular weight or a formula weight, so molecular compounds can have both. Ionic compounds must be expressed in formula weights only.

7.3 <u>Avogadro's number</u> is the number of atoms in exactly 12 g of carbon-12. This number is 6.022137×10^{23}. Therefore, Avogadro's number of chlorine atoms is 6.022137×10^{23} chlorine atoms, and Avogadro's number of diatomic chlorine molecules is 6.022137×10^{23} chlorine molecules.

7.4 A <u>mole</u> is the quantity of a substance that contains 6.022×10^{23} atoms, molecules, or formula units. The <u>molar mass</u> of a substance is the mass of one mole of the substance, in grams. The molar mass of lithium is 6.941 g. The molar mass of oxygen, O_2, is 32.00 g.

62 ■ CHAPTER 7

7.5 One mole of diatomic oxygen (O_2) contains Avogadro's number, 6.022×10^{23}, oxygen molecules. Since each oxygen molecule contains two oxygen atoms, there are $2 \times 6.022 \times 10^{23}$, or 1.204×10^{24} oxygen atoms in a mole of diatomic oxygen molecules.

7.6 One mole of ozone (O_3) contains 6.022×10^{23} ozone molecules. Since each ozone molecule contains three oxygen atoms, there are $3 \times 6.022 \times 10^{23}$, or 1.807×10^{24} oxygen atoms in a mole of ozone.

7.7 Use the relation (empirical formula)$_n$ = molecular formula. To determine n, use

$$n = \frac{\text{molecular weight}}{\text{empirical formula weight}}$$

The empirical formula of hydrogen peroxide is HO, with empirical formula weight 1.008 amu + 16.00 amu = 17.10 amu. Thus,

$$n = \frac{34 \text{ amu}}{17.01 \text{ amu}} = 2.00$$

Thus, the molecular formula of hydrogen peroxide is $(HO)_2$, or H_2O_2.

7.8 The empirical formula of $C_6H_{12}O_6$ is obtained by dividing each subscript by the greatest common factor of the subscripts. In this case, each subscript can be divided by six. This gives the empirical formula CH_2O.

■ Solutions to Practice Problems

Note on significant figures: The first time the final answer is written, it is written first with one nonsignificant figure. The least significant digit is also underlined. The final answer is then rounded to the correct number of significant figures but the least significant digit is no longer underlined. Also, in these problems, AW stands for atomic weight.

7.9 a. 2 x AW for fluorine   = 2 x 19.00 amu =   38.00 amu

b. 1 x AW for phosphorus = 1 x 30.97 amu =   30.97 amu
5 x AW for fluorine   = 5 x 19.00 amu =   95.00 amu
─────────
The molecular weight of PF_5 =   125.97 amu

CHEMICAL COMPOSITION ■ 63

c. 1 x AW for sulfur   = 1 x 32.06 amu =   32.06 amu
    3 x AW for oxygen   = 3 x 16.00 amu =   48.00 amu

               The molecular weight of SO_3 =   80.06 amu

d. 4 x AW for hydrogen = 4 x   1.008 amu =   4.032 amu
    2 x AW for carbon   = 2 x 12.01 amu =   24.02 amu
    2 x AW for oxygen   = 2 x 16.00 amu =   32.00 amu

               The molecular weight of $HC_2H_3O_2$ =   60.05̲2 amu = 60.05 amu

e. 6 x AW for carbon   = 6 x 12.01 amu =   72.06 amu
    6 x AW for hydrogen = 6 x   1.008 amu =   6.048 amu
    6 x AW for oxygen   = 6 x 16.00 amu =   96.00 amu

               The molecular weight of $C_6H_6O_6$ =   174.10̲8 amu = 174.11 amu

f. 12 x AW for carbon   = 12 x 12.01 amu = 144.12 amu
    22 x AW for hydrogen = 22 x 1.008 amu =   22.176 amu
    11 x AW for oxygen   = 11 x 16.00 amu = 176.00 amu

               The molecular weight of $C_{12}H_{22}O_{11}$ =   342.29̲6 amu = 342.30 amu

7.10 a. 2 x AW for bromine   = 2 x 79.90 amu = 159.80 amu

b. 1 x AW for sulfur   = 1 x 32.06 amu =   32.06 amu
    6 x AW for fluorine = 6 x 19.00 amu = 114.00 amu

               The molecular weight of SF_6 =   146.06 amu

c. 4 x AW for phosphorus = 4 x 30.97 amu = 123.88 amu
    10 x AW for oxygen   = 10 x 16.00 amu = 160.00 amu

               The molecular weight of P_4O_{10} =   283.88 amu = 283.88 amu

d. 2 x AW for hydrogen = 2 x   1.008 amu =   2.016 amu
    2 x AW for carbon   = 2 x 12.01 amu =   24.02 amu
    4 x AW for oxygen   = 4 x 16.00 amu =   48.00 amu

               The molecular weight of $H_2C_2O_4$ =   90.03̲6 amu = 90.04 amu

e. 2 x AW for carbon   = 2 x 12.01 amu =   24.02 amu
    4 x AW for hydrogen = 4 x   1.008 amu =   4.032 amu
    2 x AW for chlorine = 2 x 35.45 amu =   70.90 amu

               The molecular weight of $C_2H_4Cl_2$ =   98.95̲2 amu = 98.95 amu

f.   3 × AW for carbon   = 3 × 12.01 amu =   36.03 amu
     8 × AW for hydrogen = 8 ×   1.008 amu =   8.064 amu
     1 × AW for sulfur   = 1 × 32.06 amu =   32.06 amu
     ──────────
     The molecular weight of C_3H_7SH = 76.1<u>5</u>4 amu = 76.15 amu

7.11   For tartaric acid, $C_4H_6O_2$, the molecular weight is

   4 × AW for carbon   = 4 × 12.01 amu =   48.04 amu
   6 × AW for hydrogen = 6 ×   1.008 amu =   6.048 amu
   2 × AW for oxygen   = 2 × 16.00 amu =   32.00 amu
   ──────────
   86.0<u>8</u>8 amu = 86.09 amu

For glucose, $C_6H_{12}O_6$, the molecular weight is

   6 × AW for carbon   = 6 × 12.01 amu =   72.06 amu
  12 × AW for hydrogen = 12 × 1.008 amu = 12.096 amu
   6 × AW for oxygen   = 6 × 16.00 amu =   96.00 amu
   ──────────
   180.1<u>5</u>6 amu = 180.16 amu

7.12   For indigo, $C_{16}H_{10}N_2O_2$, the molecular weight is

  16 × AW for carbon   = 16 × 12.01 amu = 192.16 amu
  10 × AW for hydrogen = 10 × 1.008 amu = 10.080 amu
   2 × AW for nitrogen = 2 × 14.01 amu =   28.02 amu
   2 × AW for oxygen   = 2 × 16.00 amu =   32.00 amu
   ──────────
   262.2<u>6</u>0 amu = 262.26 amu

For indoxyl, C_8H_7ON, the molecular weight is

   8 × AW for carbon   = 8 × 12.01 amu =   96.08 amu
   7 × AW for hydrogen = 7 ×   1.008 amu =   7.056 amu
   1 × AW for oxygen   = 1 × 16.00 amu =   16.00 amu
   1 × AW for nitrogen = 1 × 14.01 amu =   14.01 amu
   ──────────
   133.1<u>4</u>6 amu = 133.15 amu

7.13 a.   1 × AW for zinc   = 1 × 65.38 amu =   65.38 amu
             2 × AW for iodine = 2 × 126.90 amu = 253.80 amu
             ──────────
             Formula weight of ZnI_2 =   319.18 amu

b.   1 x AW for aluminum =   1 x   26.98 amu =   26.98 amu
       3 x AW for bromine   =   3 x   79.90 amu =   239.70 amu

                    Formula weight of $AlBr_3$ =   266.68 amu

c.   2 x AW for carbon   =   2 x   12.01 amu =   24.02 amu
       6 x AW for hydrogen =   6 x   1.008 amu =   6.048 amu

                    Formula weight of C_2H_6 =   30.0<u>6</u>8 amu = 30.07 amu

d.   2 x AW for nitrogen   =   2 x   14.01 amu =   28.02 amu
       4 x AW for hydrogen =   4 x   1.008 amu =   4.032 amu
       3 x AW for oxygen   =   3 x   16.00 amu =   48.00 amu

                    Formula weight of NH_4NO_3 =   80.0<u>5</u>2 amu = 80.05 amu

e.   2 x AW for iron   =   2 x   55.85 amu =   111.70 amu
       3 x AW for oxygen   =   3 x   16.00 amu =   48.00 amu

                    Formula weight of Fe_2O_3 =   159.70 amu

f.   2 x AW for sodium   =   2 x   22.99 amu =   45.98 amu
       1 x AW for chromium =   1 x   52.00 amu =   52.00 amu
       4 x AW for oxygen   =   4 x   16.00 amu =   64.00 amu

                    Formula weight of Na_2CrO_4 =   161.98 amu

7.14 a.   2 x AW for potassium =   2 x   39.10 amu =   78.20 amu
           1 x AW for sulfur   =   1 x   32.06 amu =   32.06 amu
           4 x AW for oxygen   =   4 x   16.00 amu =   64.00 amu

                    Formula weight of K_2SO_4 =   174.26 amu

      b.   2 x AW for potassium =   2 x   39.10 amu =   78.20 amu
           2 x AW for chromium =   2 x   52.00 amu =   104.00 amu
           7 x AW for oxygen   =   7 x   16.00 amu =   112.00 amu

                    Formula weight of $K_2Cr_2O_7$ =   294.20 amu

      c.   3 x AW for calcium   =   3 x   40.08 amu =   120.24 amu
           2 x AW for phosphorus =   2 x   30.97 amu =   61.94 amu
           8 x AW for oxygen   =   8 x   16.00 amu =   128.00 amu

                    Formula weight of $Ca_3(PO_4)_2$ =   310.18 amu

66 ■ CHAPTER 7

d. 18 x AW for carbon   = 18 x 12.01 amu =   216.18 amu
    36 x AW for hydrogen = 36 x 1.008 amu =   36.288 amu
    18 x AW for oxygen   = 18 x 16.00 amu =   288.00 amu

    Formula weight of $C_{18}H_{36}O_{18}$ =   540.468 amu = 540.47 amu

e. 2 x AW for aluminum = 2 x 26.98 amu =   53.96 amu
    3 x AW for oxygen   = 3 x 16.00 amu =   48.00 amu

    Formula weight of Al_2O_3 =   101.96 amu

f. 1 x AW for calcium   = 1 x 40.08 amu =   40.08 amu
    1 x AW for carbon   = 1 x 12.01 amu =   12.01 amu
    3 x AW for oxygen   = 3 x 16.00 amu =   48.00 amu

    Formula weight of $CaCO_3$ =   100.09 amu

7.15 a. 117.16 amu   b. 95.21 amu   c. 259.90 amu
          d. 125.84 amu   e. 97.99 amu   f. 189.70 amu

7.16 a. 84.46 amu   b. 105.07 amu c. 265.99 amu
          d. 146.06 amu   e. 98.08 amu   f. 100.46 amu

7.17 First, convert moles to molecules using Avogadro's number.

$$2.33 \text{ mol } C_2H_5SH \times \frac{6.022 \times 10^{23} \text{ } C_2H_5SH \text{ molecules}}{1 \text{ mol } C_2H_5SH} = 1.403 \times 10^{24} \text{ molecules}$$

Next, note that one molecule of C_2H_5SH contains two atoms of carbon, seven atoms of hydrogen, and one atom of nitrogen. Using these numbers as conversion factors gives

$$1.403 \times 10^{24} \text{ molecules} \times \frac{2 \text{ C atoms}}{1 \text{ molecule}} = 2.806 \times 10^{24} = 2.81 \times 10^{24} \text{ C atoms}$$

$$1.403 \times 10^{24} \text{ molecules} \times \frac{1 \text{ S atom}}{1 \text{ molecule}} = 1.403 \times 10^{24} = 1.40 \times 10^{24} \text{ S atoms}$$

$$1.403 \times 10^{24} \text{ molecules} \times \frac{6 \text{ H atoms}}{1 \text{ molecule}} = 8.418 \times 10^{24} = 8.42 \times 10^{24} \text{ H atoms}$$

CHEMICAL COMPOSITION ■ 67

7.18 First, convert moles to molecules using Avogadro's number.

$$1.56 \text{ mol CH}_3\text{NH}_2 \times \frac{6.022 \times 10^{23} \text{ CH}_3\text{NH}_2 \text{ molecules}}{1 \text{ mol CH}_3\text{NH}_2} = 9.394 \times 10^{23} \text{ molecules}$$

Next, note that one molecule of CH_3NH_2 contains one atom of carbon, five atoms of hydrogen, and one atom of nitrogen. Using these numbers as conversion factors gives

$$9.394 \times 10^{23} \text{ molecules} \times \frac{1 \text{ C atom}}{1 \text{ molecule}} = 9.394 \times 10^{23} = 9.39 \times 10^{23} \text{ C atoms}$$

$$9.394 \times 10^{23} \text{ molecules} \times \frac{5 \text{ H atoms}}{1 \text{ molecule}} = 4.697 \times 10^{24} = 4.70 \times 10^{24} \text{ H atoms}$$

$$9.394 \times 10^{23} \text{ molecules} \times \frac{1 \text{ N atom}}{1 \text{ molecule}} = 9.394 \times 10^{23} = 9.39 \times 10^{23} \text{ N atoms}$$

7.19 First, convert moles of magnesium nitride, Mg_3N_2, to formula units using Avogadro's number.

$$0.234 \text{ mol Mg}_3\text{N}_2 \times \frac{6.022 \times 10^{23} \text{ Mg}_3\text{N}_2 \text{ formula units}}{1 \text{ mol Mg}_3\text{N}_2} = 1.409 \times 10^{23} \text{ formula units}$$

Next, note that one formula unit of Mg_3N_2 contains three magnesium ions (Mg^{2+}). Therefore,

$$1.409 \times 10^{23} \text{ formula units} \times \frac{3 \text{ Mg}^{2+} \text{ ions}}{1 \text{ formula unit}} = 4.227 \times 10^{23} = 4.23 \times 10^{23} \text{ Mg}^{2+} \text{ ions}$$

7.20 First, convert moles of calcium phosphate ($Ca_3(PO_4)_2$, to formula units using Avogadro's number.

$$3.31 \text{ mol Ca}_3(\text{PO}_4)_2 \times \frac{6.022 \times 10^{23} \text{ Ca}_3(\text{PO}_4)_2 \text{ formula units}}{1 \text{ mol Ca}_3(\text{PO}_4)_2} = 1.993 \times 10^{24} \text{ formula units}$$

Next, note that one formula unit of $Ca_3(PO_4)_2$ contains three calcium ions (Ca^{2+}). Therefore,

$$1.993 \times 10^{24} \text{ formula units} \times \frac{3 \text{ Ca}^{2+} \text{ ions}}{1 \text{ formula unit}} = 5.980 \times 10^{24} = 5.98 \times 10^{24} \text{ Ca}^{2+} \text{ ions}$$

7.21 Hydrogen is an element that exists as a diatomic molecule. Its molecular weight is

$$2 \times 1.008 \text{ amu} = 2.016 \text{ amu}$$

The molar mass of hydrogen, H_2, is 2.016 g.

Helium is an element. Its molar mass is its atomic weight expressed in grams, or 4.003 g.

7.22 Silicon is an element. Its molar mass is its atomic weight expressed in grams, or 28.09 g.

Bromine is an element that exists as a diatomic molecule. Its molecular weight is

$$2 \times 79.90 \text{ amu} = 159.80 \text{ amu}$$

The molar mass of bromine, Br_2, is 159.80 g.

7.23 Iron is an element. Its molar mass is its atomic weight expressed in grams, or 55.85 g.

Sulfur is an element that exists in S_8 molecules. Its molecular weight is

$$8 \times 32.06 \text{ amu} = 256.48 \text{ amu}$$

The molar mass of sulfur, S_8, is 256.48 g.

7.24 Calcium is an element. Its molar mass is its atomic weight expressed in grams, or 40.08 g.

Nitrogen is an element that exists as a diatomic molecule. Its molecular weight is

$$2 \times 14.01 \text{ amu} = 28.02 \text{ amu}$$

The molar mass of nitrogen, N_2, is 28.02 g.

7.25 a. The molecular weight of CH_3OH is

1 x AW for carbon	= 1 x	12.01 amu =	12.01 amu
4 x AW for hydrogen	= 4 x	1.008 amu =	4.032 amu
1 x AW for oxygen	= 1 x	16.00 amu =	16.00 amu

$$32.04\underline{2} \text{ amu} = 32.04 \text{ amu}$$

The molar mass of CH_3OH is 32.04 g.

b. The molecular weight of PF_5 is

1 x AW for phosphorus =	1 x	30.97 amu =	30.97 amu
5 x AW for fluorine =	5 x	19.00 amu =	95.00 amu
			125.97 amu

The molar mass of PF_5 is 125.97 g.

c. The molecular weight of SO_3 is

1 x AW for sulfur =	1 x	32.06 amu =	32.06 amu
3 x AW for oxygen =	3 x	16.00 amu =	48.00 amu
			80.06 amu

The molar mass of SO_3 is 80.06 g.

d. The molecular weight of $HC_2H_3O_2$ is

4 x AW for hydrogen =	4 x	1.008 amu =	4.032 amu
2 x AW for carbon =	2 x	12.01 amu =	24.02 amu
2 x AW for oxygen =	2 x	16.00 amu =	32.00 amu
			60.052 amu = 60.05 amu

The molar mass of $HC_2H_3O_2$ is 60.03 g.

e. The molecular weight of $C_6H_{12}O_6$ is

6 x AW for carbon =	6 x	12.01 amu =	72.06 amu
12 x AW for hydrogen =	12 x	1.008 amu =	12.096 amu
6 x AW for oxygen =	6 x	16.00 amu =	96.00 amu
			180.156 amu = 180.16 amu

The molar mass of $C_6H_{12}O_6$ is 180.16 g.

f. The molecular weight of $C_{12}H_{22}O_{11}$ is

12 x AW for carbon =	12 x	12.01 amu =	144.12 amu
22 x AW for hydrogen =	22 x	1.008 amu =	22.176 amu
11 x AW for oxygen =	11 x	16.00 amu =	176.00 amu
			342.296 amu = 342.30 amu

The molar mass of $C_{12}H_{22}O_{11}$ is 342.30 g.

7.26 a. The molecular weight of CH_3NH_2 is

1 × AW for carbon	=	1 ×	12.01 amu =	12.01 amu
5 × AW for hydrogen	=	5 ×	1.008 amu =	5.040 amu
1 × AW for nitrogen	=	1 ×	14.01 amu =	14.01 amu

$$31.0\underline{6}0 \text{ amu} = 31.06 \text{ amu}$$

The molar mass of CH_3NH_2 is 31.06 g.

b. The molecular weight of SF_6 is

1 × AW for sulfur	=	1 ×	32.06 amu =	32.06 amu
6 × AW for fluorine	=	6 ×	19.00 amu =	114.00 amu

$$146.06 \text{ amu}$$

The molar mass of SF_6 is 146.06 g.

c. The molecular weight of P_4O_{10} is

4 × AW for phosphorus	=	4 ×	30.97 amu =	123.88 amu
10 × AW for oxygen	=	10 ×	16.00 amu =	160.00 amu

$$283.88 \text{ amu}$$

The molar mass of P_4O_{10} is 283.88 g.

d. The molecular weight of $H_2C_2O_4$ is

2 × AW for hydrogen	=	2 ×	1.008 amu =	2.016 amu
2 × AW for carbon	=	2 ×	12.01 amu =	24.02 amu
4 × AW for oxygen	=	4 ×	16.00 amu =	64.00 amu

$$90.0\underline{3}6 \text{ amu} = 90.04 \text{ amu}$$

The molar mass of $H_2C_2O_4$ is 90.04 g.

e. The molecular weight of $C_2H_4Cl_2$ is

2 × AW for carbon	=	2 ×	12.01 amu =	24.02 amu
4 × AW for hydrogen	=	4 ×	1.008 amu =	4.032 amu
2 × AW for chlorine	=	2 ×	35.45 amu =	70.90 amu

$$98.9\underline{5}2 \text{ amu} = 98.95 \text{ amu}$$

The molar mass of $C_2H_4Cl_2$ is 98.95 g.

f. The molecular weight of C_3H_7SH is

  3 x   AW for carbon   =   3 x   12.01   amu =   36.03 amu
  8 x   AW for hydrogen   =   8 x   1.008 amu =   8.064 amu
  1 x   AW for sulfur   =   1 x   32.06   amu =   32.06 amu
                     76.1$\underline{5}$4 amu = 76.15 amu

The molar mass of C_3H_7SH is 76.15 g

7.27 a. The formula weight of LiH is

  1 x   AW for lithium   =   1 x   6.941 amu =   6.941 amu
  1 x   AW for hydrogen   =   1 x   1.008 amu =   1.008 amu
                        7.949 amu

The molar mass of LiH is 7.949 g.

b. The formula weight of $Mg(ClO_4)_2$ is

  1 x   AW for magnesium =   1 x   24.31 amu =   24.31 amu
  2 x   AW for chlorine   =   2 x   35.45 amu =   35.45 amu
  8 x   AW for oxygen   =   8 x   16.00 amu =   128.00 amu
                       223.21 amu

The molar mass of $Mg(ClO_4)_2$ is 223.21 g.

c. The formula weight of $VOCl_3$ is

  1 x   AW for vanadium   =   1 x   50.94 amu =   50.94 amu
  1 x   AW for oxygen   =   1 x   16.00 amu =   16.00 amu
  3 x   AW for chlorine   =   3 x   35.45 amu =   106.35 amu
                       173.29 amu

The molar mass of $VOCl_3$ is 173.29 g.

d. The formula weight of $Ba_3(PO_4)_2$ is

  3 x   AW for barium   =   3 x   137.33 amu =   411.99 amu
  2 x   AW for phosphorus =   2 x   30.97 amu =   61.94 amu
  8 x   AW for oxygen   =   8 x   16.00 amu =   128.00 amu
                       601.93 amu

The molar mass of $Ba_3(PO_4)_2$ is 601.93 g.

72 ■ CHAPTER 7

e. The formula weight of C_3H_7Br is

3 x AW for carbon	=	3 x	12.01 amu =	36.03 amu
7 x AW for hydrogen	=	7 x	1.008 amu =	7.056 amu
1 x AW for bromine	=	1 x	79.90 amu =	79.90 amu

122.9<u>8</u>6 amu = 122.99 amu

The molar mass of C_3H_7Br is 122.99 g.

f. The formula weight of K_2SO_3 is

2 x AW for potassium	=	2 x	39.10 amu =	78.20 amu
1 x AW for sulfur	=	1 x	32.06 amu =	32.06 amu
3 x AW for oxygen	=	3 x	16.00 amu =	48.00 amu

158.26 amu

The molar mass of K_2SO_3 is 158.26 g.

7.28 a. The formula weight of $ZnCl_2$ is

1 x AW for zinc	=	1 x	65.38 amu =	65.38 amu
2 x AW for chlorine	=	2 x	35.45 amu =	70.90 amu

136.28 amu

The molar mass of $ZnCl_2$ is 136.28 g.

b. The formula weight of $Pb(ClO_3)_2$ is

1 x AW for lead	=	1 x	207.2 amu =	207.2 amu
2 x AW for chlorine	=	2 x	35.45 amu =	70.90 amu
6 x AW for oxygen	=	6 x	16.00 amu =	96.00 amu

374.<u>1</u>0 amu = 374.1 amu

The molar mass of $Pb(ClO_3)_2$ is 374.1 g.

c. The formula weight of Al_2S_3 is

2 x AW for aluminum	=	2 x	26.98 amu =	53.96 amu
3 x AW for sulfur	=	3 x	32.06 amu =	96.18 amu

150.14 amu

The molar mass of Al_2S_3 is 150.14 g.

CHEMICAL COMPOSITION ■ 73

d. The formula weight of $(NH_4)_3PO_4$ is

3 x AW for nitrogen	= 3 x	14.01 amu =	42.03 amu
12 x AW for hydrogen	= 12 x	1.008 amu =	12.096 amu
1 x AW for phosphorus	= 1 x	30.97 amu =	30.97 amu
4 x AW for oxygen	= 4 x	16.00 amu =	64.00 amu

$\qquad\qquad\qquad\qquad\qquad\qquad\qquad\qquad$ 149.096 amu = 149.10 amu

The molar mass of $(NH_4)_3PO_4$ is 149.10 g.

e. The formula weight of C_6H_6 is

6 x AW for carbon	= 6 x	12.01 amu =	72.06 amu
6 x AW for hydrogen	= 6 x	1.008 amu =	6.048 amu

$\qquad\qquad\qquad\qquad\qquad\qquad\qquad$ 78.108 amu = 78.11 amu

The molar mass of C_6H_6 is 78.11 g.

f. The formula weight of CO_2 is

1 x AW for carbon	= 1 x 12.01 amu =	12.01 amu
2 x AW for oxygen	= 2 x 16.00 amu =	32.00 amu

$\qquad\qquad\qquad\qquad\qquad\qquad\qquad$ 44.01 amu

The molar mass of CO_2 is 44.01 g.

7.29 a. The molar mass of hydrogen, H_2, is 2.016 g. Thus,

$$1.00 \text{ g } H_2 \times \frac{1 \text{ mol } H_2}{2.016 \text{ g } H_2} = 0.4960 = 0.496 \text{ mol } H_2$$

b. The molar mass of lithium, Li, is 6.941 g. Thus,

$$1.00 \text{ g Li} \times \frac{1 \text{ mol Li}}{6.941 \text{ g Li}} = 0.1441 = 0.144 \text{ mol Li}$$

c. The molar mass of sodium, Na, is 22.99 g. Thus,

$$1.00 \text{ g Na} \times \frac{1 \text{ mol Na}}{22.99 \text{ g Na}} = 0.04350 = 0.0435 \text{ mol Na}$$

d. The molar mass of potassium, K, is 39.10 g. Thus,

$$1.00 \text{ g K} \times \frac{1 \text{ mol K}}{39.10 \text{ g K}} = 0.02558 = 0.0256 \text{ mol K}$$

74 ■ CHAPTER 7

e. The molar mass of rubidium, Rb, is 85.47 g. Thus,

$$1.00 \text{ g Rb} \times \frac{1 \text{ mol Rb}}{85.47 \text{ g Rb}} = 0.01170 = 0.0117 \text{ mol Rb}$$

f. The molar mass of cesium, Cs, is 132.91 g. Thus,

$$1.00 \text{ g Cs} \times \frac{1 \text{ mol Cs}}{132.91 \text{ g Cs}} = 7.524 \times 10^{-3} = 7.52 \times 10^{-3} \text{ mol Cs}$$

7.30 a. The molar mass of helium, He, is 4.003 g. Thus,

$$1.00 \text{ g He} \times \frac{1 \text{ mol He}}{4.003 \text{ g He}} = 0.2498 = 0.250 \text{ mol He}$$

b. The molar mass of neon, Ne, is 20.18 g. Thus,

$$1.00 \text{ g Ne} \times \frac{1 \text{ mol Ne}}{20.18 \text{ g Ne}} = 0.04955 = 0.0496 \text{ mol Ne}$$

c. The molar mass of fluorine, F_2, is 2 x 19.00 g = 38.00 g. Thus,

$$1.00 \text{ g } F_2 \times \frac{1 \text{ mol } F_2}{38.00 \text{ g } F_2} = 0.02632 = 0.0263 \text{ mol } F_2$$

d. The molar mass of iodine, I_2, is 2 x 126.90 g = 253.80 g. Thus,

$$1.00 \text{ g } I_2 \times \frac{1 \text{ mol } I_2}{253.80 \text{ g } I_2} = 3.940 \times 10^{-3} = 3.94 \times 10^{-3} \text{ mol } I_2$$

e. The molar mass of nitrogen, N_2, is 2 x 14.01 g = 28.02 g. Thus,

$$1.00 \text{ g } N_2 \times \frac{1 \text{ mol } N_2}{28.02 \text{ g } N_2} = 0.03569 = 0.0357 \text{ mol } N_2$$

f. The molar mass of phosphorus, P_4, is 4 x 30.97 g = 123.88 g. Thus,

$$1.00 \text{ g } P_4 \times \frac{1 \text{ mol } P_4}{123.88 \text{ g } P_4} = 8.072 \times 10^{-3} = 8.07 \times 10^{-3} \text{ mol } P_4$$

7.31 Note: See problem #25 for the molar mass calculations for this problem.

a. The molar mass of CH_3OH is 32.04 g. Thus,

$$5.00 \text{ g } CH_3OH \times \frac{1 \text{ mol } CH_3OH}{32.04 \text{ g } CH_3OH} = 0.1561 = 0.156 \text{ mol } CH_3OH$$

b. The molar mass of PF_5 is 125.97 g. Thus,

$$5.00 \text{ g } PF_5 \times \frac{1 \text{ mol } PF_5}{125.97 \text{ g } PF_5} = 0.03969 = 0.0397 \text{ mol } PF_5$$

c. The molar mass of SO_3 is 80.06 g. Thus,

$$5.00 \text{ g } SO_3 \times \frac{1 \text{ mol } SO_3}{80.06 \text{ g } SO_3} = 0.06245 = 0.0625 \text{ mol } SO_3$$

d. The molar mass of $HC_2H_3O_2$ is 60.03 g. Thus,

$$5.00 \text{ g } HC_2H_3O_2 \times \frac{1 \text{ mol } HC_2H_3O_2}{60.03 \text{ g } HC_2H_3O_2} = 0.08329 = 0.0833 \text{ mol } HC_2H_3O_2$$

e. The molar mass of $C_6H_{12}O_6$ is 180.16 g. Thus,

$$5.00 \text{ g } C_6H_{12}O_6 \times \frac{1 \text{ mol } C_6H_{12}O_6}{180.16 \text{ g } C_6H_{12}O_6} = 0.02775 = 0.0278 \text{ mol } C_6H_{12}O_6$$

f. The molar mass of $C_{12}H_{22}O_{11}$ is 342.30 g. Thus,

$$5.00 \text{ g } C_{12}H_{22}O_{11} \times \frac{1 \text{ mol } C_{12}H_{22}O_{11}}{342.30 \text{ g } C_{12}H_{22}O_{11}} = 0.01461 = 0.0146 \text{ mol } C_{12}H_{22}O_{11}$$

7.32 Note: See problem #26 for the molar mass calculations for this problem.

a. The molar mass of CH_3NH_2 is 31.06 g. Thus,

$$5.00 \text{ g } CH_3NH_2 \times \frac{1 \text{ mol } CH_3NH_2}{31.06 \text{ g } CH_3NH_2} = 0.1610 = 0.161 \text{ mol } CH_3NH_2$$

b. The molar mass of SF_6 is 146.06 g. Thus,

$$5.00 \text{ g } SF_6 \times \frac{1 \text{ mol } SF_6}{146.06 \text{ g } SF_6} = 0.03423 = 0.0342 \text{ mol } SF_6$$

c. The molar mass of P_4O_{10} is 283.88 g. Thus,

$$5.00 \text{ g } P_4O_{10} \times \frac{1 \text{ mol } P_4O_{10}}{283.88 \text{ g } P_4O_{10}} = 0.01761 = 0.0176 \text{ mol } P_4O_{10}$$

d. The molar mass of $H_2C_2O_4$ is 90.04 g. Thus,

$$5.00 \text{ g } H_2C_2O_4 \times \frac{1 \text{ mol } H_2C_2O_4}{90.04 \text{ g } H_2C_2O_4} = 0.05553 = 0.0555 \text{ mol } H_2C_2O_4$$

e. The molar mass of $C_2H_4Cl_2$ is 98.95 g. Thus,

$$5.00 \text{ g } C_2H_4Cl_2 \times \frac{1 \text{ mol } C_2H_4Cl_2}{98.95 \text{ g } C_2H_4Cl_2} = 0.05053 = 0.0505 \text{ mol } C_2H_4Cl_2$$

f. The molar mass of C_3H_7SH is 76.15 g. Thus,

$$5.00 \text{ g } C_3H_7SH \times \frac{1 \text{ mol } C_3H_7SH}{76.15 \text{ g } C_3H_7SH} = 0.06566 = 0.0657 \text{ mol } C_3H_7SH$$

7.33 The molecular weight of hydrogen sulfide, H_2S, is

2 x AW for hydrogen	=	2 x	1.008 amu =	2.016 amu
1 x AW for sulfur	=	1 x	32.06 amu =	32.06 amu
				34.076 amu

Therefore, the molar mass of H_2S is 34.076 g. Thus,

$$1.11 \times 10^{-2} \text{ mol } H_2S \times \frac{34.076 \text{ g } H_2S}{1 \text{ mol } H_2S} = 0.3782 = 0.378 \text{ g } H_2S$$

7.34 The molecular weight of dinitrogen monoxide, N_2O, is

2 x AW for nitrogen	=	2 x	14.01 amu =	28.02 amu
1 x AW for oxygen	=	1 x	16.00 amu =	16.00 amu
				44.02 amu

Therefore, the molar mass of N_2O is 44.02 g. Thus,

$$3.45 \times 10^{-4} \text{ mol } N_2O \times \frac{44.02 \text{ g } N_2O}{1 \text{ mol } N_2O} = 0.01519 = 0.0152 \text{ g } N_2O$$

7.35 The molar mass of H⁺ ions is 1.008 g. Thus,

$$1.3 \times 10^{-3} \text{ mol H}^+ \times \frac{1.008 \text{ g H}^+}{1 \text{ mol H}^+} = 1.31 \times 10^{-3} = 1.3 \times 10^{-3} \text{ g H}^+$$

7.36 The formula weight of a phosphate ion, PO_4^{3-}, is

1 × AW for phosphorus =	1 ×	30.97 amu =	30.97 amu
4 × AW for oxygen =	4 ×	16.00 amu =	64.00 amu
			94.97 amu

Therefore, the molar mass of PO_4^{3-} ions is 94.97 g. Thus,

$$6.2 \times 10^{-2} \text{ mol PO}_4^{3-} \times \frac{94.97 \text{ g PO}_4^{3-}}{1 \text{ mol PO}_4^{3-}} = 5.89 = 5.9 \text{ g PO}_4^{3-}$$

7.37 First, find the molar mass of $H_2C_2O_4$. The molecular weight is

2 × AW for hydrogen =	2 ×	1.008 amu =	2.016 amu
2 × AW for carbon =	2 ×	12.01 amu =	24.02 amu
4 × AW for oxygen =	4 ×	16.00 amu =	64.00 amu
			90.036 amu

Thus, the molar mass of $H_2C_2O_4$ is 90.036 g. This mass of oxalic acid contains 2.016 g of hydrogen, 24.02 g of carbon, and 64.00 g of oxygen. Using these numbers as conversion factors gives the mass percentages.

$$\text{mass percent of hydrogen} = \frac{2.016 \text{ g}}{90.036 \text{ g}} \times 100\% = 2.2391\% = 2.239\%$$

$$\text{mass percent of carbon} = \frac{24.02 \text{ g}}{90.036 \text{ g}} \times 100\% = 26.678\% = 26.68\%$$

$$\text{mass percent of oxygen} = \frac{64.00 \text{ g}}{90.036 \text{ g}} \times 100\% = 71.082\% = 71.08\%$$

7.38 First, find the molar mass of $C_6H_{10}O_4$. The molecular weight is

6 × AW for carbon	= 6 × 12.01 amu =	72.06 amu
10 × AW for hydrogen	= 10 × 1.008 amu =	10.08 amu
4 × AW for oxygen	= 4 × 16.00 amu =	64.00 amu
		146.14 amu

Thus, the molar mass of $C_6H_{10}O_4$ is 146.14 g. This mass of adipic acid contains 72.06 g of carbon, 10.08 g of hydrogen, and 64.00 g of oxygen. Using these numbers as conversion factors gives the mass percentages.

mass percent of carbon = $\dfrac{72.06 \text{ g}}{146.14 \text{ g}}$ × 100 % = 49.309 % = 49.31 %

mass percent of hydrogen = $\dfrac{10.08 \text{ g}}{146.14 \text{ g}}$ × 100 % = 6.8975 % = 6.898 %

mass percent of oxygen = $\dfrac{64.00 \text{ g}}{146.14 \text{ g}}$ × 100 % = 43.794 % = 43.79 %

7.39 First, find the molar mass of $C_6H_4Cl_2$. The molecular weight is

6 × AW for carbon	= 6 × 12.01 amu =	72.06 amu
4 × AW for hydrogen	= 4 × 1.008 amu =	4.032 amu
2 × AW for chlorine	= 2 × 35.45 amu =	70.90 amu
		146.992 amu

Thus, the molar mass of $C_6H_4Cl_2$ is 146.992 g. This mass of para-dichlorobenzene contains 72.06 g of carbon, 4.032 g of hydrogen, and 70.90 g of chlorine. Using these numbers as conversion factors gives the mass percentages.

mass percent of carbon = $\dfrac{72.06 \text{ g}}{146.992 \text{ g}}$ × 100 % = 49.023 % = 49.02 %

mass percent of hydrogen = $\dfrac{4.032 \text{ g}}{146.992 \text{ g}}$ × 100 % = 2.7430 % = 2.743 %

mass percent of chlorine = $\dfrac{70.90 \text{ g}}{146.992 \text{ g}}$ × 100 % = 48.234 % = 48.23 %

7.40 First, find the molar mass of $C_6H_8O_2$. The molecular weight is

      6 x AW for carbon   = 6 x 12.01 amu =   72.06 amu
      8 x AW for hydrogen = 8 x 1.008 amu =   8.064 amu
      2 x AW for oxygen   = 2 x 16.00 amu = 32.00 amu
                                                                     112.124 amu

Thus, the molar mass of $C_6H_8O_2$ is 112.124 g. This mass of sorbic acid contains 72.06 g of carbon, 8.064 g of hydrogen, and 32.00 g of oxygen. Using these numbers as conversion factors gives the mass percentages.

$$\text{mass percent of carbon} = \frac{72.06 \text{ g}}{112.124 \text{ g}} \times 100\% = 64.268\% = 64.27\%$$

$$\text{mass percent of hydrogen} = \frac{8.064 \text{ g}}{112.124 \text{ g}} \times 100\% = 7.1920\% = 7.192\%$$

$$\text{mass percent of oxygen} = \frac{32.00 \text{ g}}{112.124 \text{ g}} \times 100\% = 28.540\% = 28.54\%$$

7.41 First, find the mass percentage of chlorine in CsCl. The formula weight is

      1 x AW for cesium   = 1 x 132.91 amu = 132.91 amu
      1 x AW for chlorine = 1 x 35.45 amu =   35.45 amu
                                                                     168.36 amu

Thus, the molar mass of CsCl is 168.36 g. This mass of cesium chloride contains 35.45 g of chlorine. Thus, the mass percentage of chlorine is

$$\text{mass percent of chlorine in CsCl} = \frac{35.45 \text{ g}}{168.36 \text{ g}} \times 100\% = 21.056\% = 21.06\%$$

80 ■ CHAPTER 7

Repeat the calculation for NaCl. The formula weight is

1 x AW for sodium =	1 x 22.99 amu =	22.99 amu
1 x AW for chlorine =	1 x 35.45 amu =	35.45 amu
		58.44 amu

Thus, the molar mass of NaCl is 58.44 g. This mass of sodium chloride contains 35.45 g of chlorine. Thus, the mass percentage of chlorine is

mass percent of chlorine in NaCl = $\dfrac{35.45\ g}{58.44\ g}$ x 100 % = 60.661 % = 60.66 %

Since NaCl has a higher mass percentage of chlorine, it contains more chlorine per gram of compound.

7.42 First, find the mass percentage of sodium in Na_2CO_3. The formula weight is

2 x AW for sodium =	2 x 22.99 amu =	45.98 amu
1 x AW for carbon =	1 x 12.01 amu =	12.01 amu
3 x AW for oxygen =	3 x 16.00 amu =	48.00 amu
		105.99 amu

Thus, the molar mass of Na_2CO_3 is 105.99 g. This mass of sodium carbonate contains 45.98 g of sodium. Thus, the mass percentage of sodium is

mass percent of sodium in Na_2CO_3 = $\dfrac{45.98\ g}{105.99\ g}$ x 100 % = 43.382 % = 43.38 %

Repeat the calculation for NaCl. The formula weight is

1 x AW for sodium =	1 x 22.99 amu =	22.99 amu
1 x AW for chlorine =	1 x 35.45 amu =	35.45 amu
		58.44 amu

Thus, the molar mass of NaCl is 58.44 g. This mass of sodium chloride contains 22.99 g of sodium. Thus, the mass percentage of sodium is

mass percent of sodium in NaCl = $\dfrac{22.99\ g}{58.44\ g}$ x 100% = 39.339 % = 39.34 %

Since Na_2CO_3 has a higher mass percentage of sodium, it contains more sodium per gram of compound.

CHEMICAL COMPOSITION ■ 81

7.43 In 100 g of the compound there is 43.64 g of phosphorus and 56.36 g of oxygen. Convert these quantities to moles using the molar masses as conversion factors.

mol phosphorus = 43.64 g P × $\dfrac{1 \text{ mol P}}{30.97 \text{ g P}}$ = 1.409 mol P

mol oxygen = 56.36 g O × $\dfrac{1 \text{ mol O}}{16.00 \text{ g O}}$ = 3.523 mol O

Divide both quantities by the smallest number of moles to get the subscripts.

subscript for phosphorus = $\dfrac{1.409 \text{ mol}}{1.409 \text{ mol}}$ = 1.000

subscript for oxygen = $\dfrac{3.523 \text{ mol}}{1.409 \text{ mol}}$ = 2.500

The formula is now $P_{1.000}O_{2.500}$. You need to multiply both subscripts by two to convert of whole numbers. The empirical formula is P_2O_5.

7.44 In 100 g of the compound there is 36.84 g of nitrogen and 63.16 g of oxygen. Convert these quantities to moles using the molar masses as conversion factors.

mol nitrogen = 36.84 g N × $\dfrac{1 \text{ mol N}}{14.01 \text{ g N}}$ = 2.630 mol N

mol oxygen = 63.16 g O × $\dfrac{1 \text{ mol O}}{16.00 \text{ g O}}$ = 3.948 mol O

Divide both quantities by the smallest number of moles to get the subscripts.

subscript for nitrogen = $\dfrac{2.630 \text{ mol}}{2.630 \text{ mol}}$ = 1.000

subscript for oxygen = $\dfrac{3.948 \text{ mol}}{2.630 \text{ mol}}$ = 1.501

The formula is now $N_{1.000}O_{1.501}$. You need to multiply both subscripts by two to convert to whole numbers. The empirical formula is N_2O_3.

7.45 In 100 g of the compound there is 65.4 g of carbon, 5.5 g of hydrogen, and 29.1 g of oxygen. Convert these quantities to moles using the molar masses as conversion factors.

$$\text{mol carbon} = 65.4 \text{ g C} \times \frac{1 \text{ mol C}}{12.01 \text{ g C}} = 5.45 \text{ mol C}$$

$$\text{mol hydrogen} = 5.5 \text{ g H} \times \frac{1 \text{ mol H}}{1.008 \text{ g H}} = 5.5 \text{ mol H}$$

$$\text{mol oxygen} = 29.1 \text{ g O} \times \frac{1 \text{ mol O}}{16.00 \text{ g O}} = 1.82 \text{ mol O}$$

Divide each quantity by the smallest number of moles to get the subscripts.

$$\text{subscript for carbon} = \frac{5.45 \text{ mol}}{1.82 \text{ mol}} = 2.99$$

$$\text{subscript for hydrogen} = \frac{5.5 \text{ mol}}{1.82 \text{ mol}} = 3.0$$

$$\text{subscript for oxygen} = \frac{1.82 \text{ mol}}{1.82 \text{ mol}} = 1.00$$

Since all of the subscripts are whole numbers, the empirical formula is C_3H_3O.

7.46 In 100 g of the compound there is 34.6 g of carbon, 3.9 g of hydrogen, and 61.5 g of oxygen. Convert these quantities to moles using the molar masses as conversion factors.

$$\text{mol carbon} = 34.6 \text{ g C} \times \frac{1 \text{ mol C}}{12.01 \text{ g C}} = 2.88 \text{ mol C}$$

$$\text{mol hydrogen} = 3.9 \text{ g H} \times \frac{1 \text{ mol H}}{1.008 \text{ g H}} = 3.9 \text{ mol H}$$

$$\text{mol oxygen} = 61.5 \text{ g O} \times \frac{1 \text{ mol O}}{16.00 \text{ g O}} = 3.84 \text{ mol O}$$

CHEMICAL COMPOSITION ■ 83

Divide each quantity by the smallest number of moles to get the subscripts.

subscript for carbon = $\dfrac{2.88 \text{ mol}}{2.88 \text{ mol}}$ = 1.00

subscript for hydrogen = $\dfrac{3.9 \text{ mol}}{2.88 \text{ mol}}$ = 1.4

subscript for oxygen = $\dfrac{3.84 \text{ mol}}{2.88 \text{ mol}}$ = 1.33

The formula is now $C_{1.00}H_{1.35}O_{1.33}$. You must multiply each subscript by three to convert to whole numbers. The empirical formula is $C_3H_4O_4$.

7.47 The empirical formula weight of CHCl is

$$
\begin{array}{llll}
1 \times \text{AW for carbon} & = 1 \times & 12.01 \text{ amu} = & 12.01 \text{ amu} \\
1 \times \text{AW for hydrogen} & = 1 \times & 1.008 \text{ amu} = & 1.008 \text{ amu} \\
1 \times \text{AW for chlorine} & = 1 \times & 35.45 \text{ amu} = & 35.45 \text{ amu} \\
& & & \overline{46.468 \text{ amu}}
\end{array}
$$

Find the value of n

$n = \dfrac{\text{molecular weight}}{\text{empirical formula weight}} = \dfrac{291 \text{ amu}}{48.468 \text{ amu}} = 6.00$

The molecular formula is $(CHCl)_6$, or $C_6H_6Cl_6$.

7.48 The empirical formula weight of C_2H_6N is

$$
\begin{array}{llll}
2 \times \text{AW for carbon} & = 2 \times & 12.01 \text{ amu} = & 24.02 \text{ amu} \\
6 \times \text{AW for hydrogen} & = 6 \times & 1.008 \text{ amu} = & 6.048 \text{ amu} \\
1 \times \text{AW for nitrogen} & = 1 \times & 14.01 \text{ amu} = & 14.01 \text{ amu} \\
& & & \overline{44.078 \text{ amu}}
\end{array}
$$

Find the value of n

$n = \dfrac{\text{molecular weight}}{\text{empirical formula weight}} = \dfrac{88 \text{ amu}}{44.078 \text{ amu}} = 2.0$

The molecular formula is $(C_2H_6N)_2$, or $C_4H_{12}N_2$.

84 ■ CHAPTER 7

7.49 The empirical formula weight of P_2O_5 is

$$2 \times \text{AW for phosphorus} = 2 \times 30.97 \text{ amu} = 61.94 \text{ amu}$$
$$5 \times \text{AW for oxygen} = 5 \times 16.00 \text{ amu} = 80.00 \text{ amu}$$
$$\overline{141.94 \text{ amu}}$$

Find the value of n

$$n = \frac{\text{molecular weight}}{\text{empirical formula weight}} = \frac{284 \text{ amu}}{141.94 \text{ amu}} = 2.00$$

The molecular formula is $(P_2O_5)_2$, or P_4O_{10}.

7.50 The empirical formula weight of NO_2 is

$$1 \times \text{AW for nitrogen} = 1 \times 14.01 \text{ amu} = 14.01 \text{ amu}$$
$$2 \times \text{AW for oxygen} = 2 \times 16.00 \text{ amu} = 32.00 \text{ amu}$$
$$\overline{46.01 \text{ amu}}$$

Find the value of n

$$n = \frac{\text{molecular weight}}{\text{empirical formula weight}} = \frac{46 \text{ amu}}{46.01 \text{ amu}} = 1.0$$

The molecular formula is $(NO_2)_1$, or NO_2.

7.51 In 100 g of the substance there is 26.7 g of carbon, 2.2 g of hydrogen, and 71.7 g of oxygen. Convert these quantities to moles using the molar masses as conversion factors.

$$\text{mol carbon} = 26.7 \text{ g C} \times \frac{1 \text{ mol C}}{12.01 \text{ g C}} = 2.22 \text{ mol C}$$

$$\text{mol hydrogen} = 2.2 \text{ g H} \times \frac{1 \text{ mol H}}{1.008 \text{ g H}} = 2.2 \text{ mol H}$$

$$\text{mol oxygen} = 71.1 \text{ g O} \times \frac{1 \text{ mol O}}{16.00 \text{ g O}} = 4.44 \text{ mol O}$$

CHEMICAL COMPOSITION ■ 85

Divide each quantity by the smallest number of moles to get the subscripts.

subscript for carbon = $\dfrac{2.22 \text{ mol}}{2.2 \text{ mol}}$ = 1.0

subscript for hydrogen = $\dfrac{2.2 \text{ mol}}{2.2 \text{ mol}}$ = 1.0

subscript for oxygen = $\dfrac{4.44 \text{ mol}}{2.2 \text{ mol}}$ = 2.0

The subscripts are all whole numbers, so the empirical formula is CHO_2. The empirical formula weight is

1 x AW for carbon	=	1 x	12.01 amu	=	12.01 amu
1 x AW for hydrogen	=	1 x	1.008 amu	=	1.008 amu
2 x AW for oxygen	=	2 x	16.00 amu	=	32.00 amu
					45.018 amu

Find the value of n

$n = \dfrac{\text{molecular weight}}{\text{empirical formula weight}} = \dfrac{90 \text{ amu}}{45.018 \text{ amu}} = 2.0$

Thus, the molecular formula is $(CHO_2)_2$, or $C_2H_2O_4$.

7.52 In 100 g of the substance there is 49.3 g of carbon, 6.9 g of hydrogen, and 43.8 g of oxygen. Convert these quantities to moles using the molar masses as conversion factors.

mol carbon = 49.3 g C x $\dfrac{1 \text{ mol C}}{12.01 \text{ g C}}$ = 4.10 mol C

mol hydrogen = 6.9 g H x $\dfrac{1 \text{ mol H}}{1.008 \text{ g H}}$ = 6.8 mol H

mol oxygen = 43.8 g O x $\dfrac{1 \text{ mol O}}{16.00 \text{ g O}}$ = 2.74 mol O

Divide each quantity by the smallest number of moles to get the subscripts.

subscript for carbon = $\dfrac{4.10 \text{ mol}}{2.74 \text{ mol}}$ = 1.50

subscript for hydrogen = $\dfrac{6.8 \text{ mol}}{2.74 \text{ mol}}$ = 2.5

subscript for oxygen = $\dfrac{2.74 \text{ mol}}{2.74 \text{ mol}}$ = 1.00

The formula is now $C_{1.50}H_{2.5}O_{1.00}$. You must multiply each subscript by two to convert to whole numbers. The empirical formula is $C_3H_5O_2$. The empirical formula weight is

```
3 x AW for carbon    = 3 x  12.01 amu =   36.03 amu
5 x AW for hydrogen  = 5 x   1.008 amu =   5.040 amu
2 x AW for oxygen    = 2 x  16.00 amu =   32.00 amu
                                         ─────────
                                          73.070 amu
```

Find the value of n

n = $\dfrac{\text{molecular weight}}{\text{empirical formula weight}}$ = $\dfrac{146 \text{ amu}}{73.070 \text{ amu}}$ = 2.00

The molecular formula is $(C_3H_5O_2)_2$, or $C_6H_{10}O_4$.

7.53 In 100 g of the substance there is 49.0 g of carbon, 2.7 g of hydrogen, and 48.2 g of chlorine. Convert these quantities to moles using the molar masses as conversion factors.

mol carbon = 49.0 g C × $\dfrac{1 \text{ mol C}}{12.01 \text{ g C}}$ = 4.08 mol C

mol hydrogen = 2.7 g H × $\dfrac{1 \text{ mol H}}{1.008 \text{ g H}}$ = 2.7 mol H

mol chlorine = 48.2 g Cl × $\dfrac{1 \text{ mol Cl}}{35.45 \text{ g Cl}}$ = 1.36 mol Cl

CHEMICAL COMPOSITION ■ 87

Divide each quantity by the smallest number of moles to get the subscripts.

subscript for carbon = $\dfrac{4.08 \text{ mol}}{1.36 \text{ mol}}$ = 3.00

subscript for hydrogen = $\dfrac{2.7 \text{ mol}}{1.36 \text{ mol}}$ = 2.0

subscript for chlorine = $\dfrac{1.36 \text{ mol}}{1.36 \text{ mol}}$ = 1.00

Since all of the subscripts are whole numbers, the empirical formula is C_3H_2Cl. The empirical formula weight is

```
3 x AW for carbon    = 3 x 12.01 amu =   36.03 amu
2 x AW for hydrogen  = 2 x  1.008 amu =   2.016 amu
1 x AW for chlorine  = 1 x 35.45 amu =   35.45 amu
                                         ─────────
                                         73.496 amu
```

Find the value of n

$n = \dfrac{\text{molecular weight}}{\text{empirical formula weight}} = \dfrac{147 \text{ amu}}{73.496 \text{ amu}} = 2.00$

The molecular formula is $(C_3H_2Cl)_2$, or $C_6H_4Cl_2$.

7.54 In 100 g of the substance there is 64.3 g of carbon, 7.2 g of hydrogen, and 28.5 g of oxygen. Convert these quantities to moles using the molar masses as conversion factors.

mol carbon = 64.3 g C × $\dfrac{1 \text{ mol C}}{12.01 \text{ g C}}$ = 5.35 mol C

mol hydrogen = 7.2 g H × $\dfrac{1 \text{ mol H}}{1.008 \text{ g H}}$ = 7.1 mol H

mol oxygen = 28.5 g O × $\dfrac{1 \text{ mol O}}{16.00 \text{ g O}}$ = 1.78 mol O

88 ■ CHAPTER 7

Divide each quantity by the smallest number of moles to get the subscripts.

$$\text{subscript for carbon} = \frac{5.35 \text{ mol}}{1.78 \text{ mol}} = 3.01$$

$$\text{subscript for hydrogen} = \frac{7.1 \text{ mol}}{1.78 \text{ mol}} = 4.0$$

$$\text{subscript for oxygen} = \frac{1.78 \text{ mol}}{1.78 \text{ mol}} = 1.00$$

Since all of the subscripts are whole numbers, the empirical formula is C_3H_4O. The empirical formula weight is

3 x AW for carbon	=	3 x 12.01 amu	=	36.03 amu
4 x AW for hydrogen	=	4 x 1.008 amu	=	4.032 amu
1 x AW for oxygen	=	1 x 16.00 amu	=	16.00 amu
				56.062 amu

Find the value of n

$$n = \frac{\text{molecular weight}}{\text{empirical formula weight}} = \frac{112 \text{ amu}}{56.062 \text{ amu}} = 2.00$$

The molecular formula is $(C_3H_4O)_2$, or $C_6H_8O_2$.

■ Solutions to Additional Problems

Note on significant figures: The first time the final answer is written, it is written first with one nonsignificant figure. The least significant digit is also underlined. The final answer is then rounded to the correct number of significant figures but the least significant digit is no longer underlined. Also, in these problems, AW stands for atomic weight.

7.55 $6.3 \text{ g carbon-12} \times \dfrac{1 \text{ mol carbon-12}}{12.00 \text{ g carbon-12}} = 0.52\underline{5} = 0.53 \text{ mol carbon-12}$

$0.525 \text{ mol carbon-12} \times \dfrac{6.022 \times 10^{23} \text{ atoms}}{1 \text{ mol carbon-12}} = 3.1\underline{6} \times 10^{23} = 3.2 \times 10^{23} \text{ atoms}$

CHEMICAL COMPOSITION ◼ 89

7.56 $0.010 \text{ g carbon-12} \times \dfrac{1 \text{ mol carbon-12}}{12.00 \text{ g carbon-12}} = 8.33 \times 10^{-4} = 8.3 \times 10^{-4} \text{ mol carbon-12}$

$8.33 \times 10^{-4} \text{ mol carbon-12} \times \dfrac{6.022 \times 10^{23} \text{ atoms}}{1 \text{ mol carbon-12}} = 5.02 \times 10^{20} = 5.0 \times 10^{20} \text{ atoms}$

7.57 $2.43 \times 10^{24} \text{ atoms} \times \dfrac{1 \text{ mol carbon-12}}{6.022 \times 10^{23} \text{ atoms}} \times \dfrac{12.00 \text{ g}}{1 \text{ mol carbon-12}} = 48.42 = 48.4 \text{ g}$

7.58 $8.32 \times 10^{23} \text{ atoms} \times \dfrac{1 \text{ mol carbon-12}}{6.022 \times 10^{23} \text{ atoms}} \times \dfrac{12.00 \text{ g}}{1 \text{ mol carbon-12}} = 16.58 = 16.6 \text{ g}$

7.59 $3.29 \times 10^{-2} \text{ mol Na} \times \dfrac{22.99 \text{ g}}{1 \text{ mol Na}} \times \dfrac{1000 \text{ mg}}{1 \text{ g}} = 756.4 = 756 \text{ mg}$

7.60 The molar mass of calcium carbonate, $CaCO_3$, is

1 x AW for calcium	=	1 x 40.08 amu =	40.08 amu
1 x AW for carbon	=	1 x 12.01 amu =	12.01 amu
3 x AW for oxygen	=	3 x 16.00 amu =	48.00 amu
			100.09 amu

The molar mass of $CaCO_3$ is 100.09 g. Therefore,

$2.02 \times 10^{-3} \text{ mol CaCO}_3 \times \dfrac{100.09 \text{ g}}{1 \text{ mol CaCO}_3} \times \dfrac{1000 \text{ mg}}{1 \text{ g}} = 202.2 = 202 \text{ mg}$

7.61 The formula weight of magnesium chloride, $MgCl_2$, is

1 x AW for magnesium	=	1 x 24.31 amu =	24.31 amu
2 x AW for chlorine	=	2 x 35.45 amu =	70.90 amu
			95.21 amu

Thus, the molar mass of $MgCl_2$ is 95.21 g.

Since there are two chloride ions per formula unit, you get

$$1.11 \text{ kg MgCl}_2 \times \frac{1000 \text{ g}}{1 \text{ kg}} \times \frac{1 \text{ mol MgCl}_2}{95.21 \text{ g MgCl}_2} \times \frac{2 \text{ mol Cl}^-}{1 \text{ mol MgCl}_2} = 23.32 = 23.3 \text{ mol Cl}^-$$

$$23.32 \text{ mol Cl}^- \times \frac{6.022 \times 10^{23} \text{ Cl}^- \text{ ions}}{1 \text{ mol Cl}^- \text{ ions}} = 1.404 \times 10^{25} = 1.40 \times 10^{25} \text{ Cl}^- \text{ ions}$$

7.62 The formula weight of sodium phosphate, Na_3PO_4, is

3 x AW for sodium	=	3 x	22.99 amu =	68.97 amu
1 x AW for phosphorus	=	1 x	30.97 amu =	30.97 amu
4 x AW for oxygen	=	4 x	16.00 amu =	64.00 amu
				163.94 amu

Thus, the molar mass of Na_3PO_4 is 163.94 g. Since there are three sodium ions per formula unit, you get

$$2.452 \text{ kg Na}_3\text{PO}_4 \times \frac{1000 \text{ g}}{1 \text{ kg}} \times \frac{1 \text{ mol Na}_3\text{PO}_4}{163.94 \text{ g Na}_3\text{PO}_4} \times \frac{3 \text{ mol Na}^+}{1 \text{ mol Na}_3\text{PO}_4}$$

$$= 44.870 = 44.87 \text{ mol Na}^+$$

$$44.870 \text{ mol Na}^+ \times \frac{6.022 \times 10^{23} \text{ Na}^+ \text{ ions}}{1 \text{ mol Na}^+ \text{ ions}} = 2.7021 \times 10^{25} = 2.702 \times 10^{25} \text{ Na}^+ \text{ ions}$$

7.63 Since there are eight sulfur atoms per S_8 molecule, you get

$$6.022 \times 10^{23} \text{ S atoms} \times \frac{1 \text{ S}_8 \text{ molecule}}{8 \text{ S atoms}} = 7.5275 \times 10^{22} = 7.528 \times 10^{22} \text{ S}_8 \text{ molecules}$$

7.64 Since there are four phosphorus atoms per P_4 molecule, you get

$$6.022 \times 10^{23} \text{ P atoms} \times \frac{1 \text{ P}_4 \text{ molecule}}{4 \text{ P atoms}} = 1.5055 \times 10^{22} = 1.506 \times 10^{23} \text{ P}_4 \text{ molecules}$$

CHEMICAL COMPOSITION ■ 91

7.65 The formula weight of calcium carbonate, $CaCO_3$, is

    1 x AW for calcium   = 1 x 40.08 amu =   40.08 amu
    1 x AW for carbon   = 1 x 12.01 amu =   12.01 amu
    3 x AW for oxygen   = 3 x 16.00 amu =   48.00 amu
                                                                              100.09 amu

Thus, the molar mass of $CaCO_3$ is 100.09 g. Since there is one calcium ion per formula unit, you get

$$3.4 \text{ g CaCO}_3 \times \frac{1 \text{ mol CaCO}_3}{100.09 \text{ g CaCO}_3} \times \frac{1 \text{ mol Ca}^{2+} \text{ ions}}{1 \text{ mol CaCO}_3} \times \frac{6.022 \times 10^{23} \text{ Ca}^{2+} \text{ ions}}{1 \text{ mol Ca}^{2+} \text{ ions}}$$

$$= 2.04 \times 10^{22} = 2.0 \times 10^{22} \text{ Ca}^{2+} \text{ ions}$$

7.66 The formula weight of lithium chloride, LiCl, is

    1 x AW for lithium   = 1 x 6.941 amu =   6.941 amu
    1 x AW for chlorine = 1 x 35.45 amu = 35.45 amu
                                                                          42.391 amu

Thus, the molar mass of LiCl is 42.391 g. Since there is one chloride ion per formula unit, you get

$$11.1 \text{ g LiCl} \times \frac{1 \text{ mol LiCl}}{42.391 \text{ g LiCl}} \times \frac{1 \text{ mol Cl}^- \text{ ions}}{1 \text{ mol LiCl}} \times \frac{6.022 \times 10^{23} \text{ Cl}^- \text{ ions}}{1 \text{ mol Cl}^- \text{ ions}}$$

$$= 1.577 \times 10^{23} = 1.58 \times 10^{23} \text{ Cl}^- \text{ ions}$$

7.67 The molecular weight of water, H_2O, is

    2 x AW for hydrogen = 2 x 1.008 amu =   2.016 amu
    1 x AW for oxygen   = 1 x 16.00 amu = 16.00 amu
                                                                          18.016 amu

Thus, the molar mass of water is 18.016 g. Since there are two hydrogen atoms per molecule, you get

$$2.4 \text{ g H}_2\text{O} \times \frac{1 \text{ mol H}_2\text{O}}{18.016 \text{ g H}_2\text{O}} \times \frac{6.022 \times 10^{23} \text{ H}_2\text{O molecules}}{1 \text{ mol H}_2\text{O}} \times \frac{2 \text{ H atoms}}{1 \text{ H}_2\text{O molecule}}$$

$$= 1.60 \times 10^{23} = 1.6 \times 10^{23} \text{ H atoms}$$

92 ■ CHAPTER 7

7.68 The molecular weight of iodine pentafluoride, IF_5, is

$$
\begin{array}{lllll}
1 \times \text{AW for iodine} & = & 1 \times 126.90 \text{ amu} = & 126.90 \text{ amu} \\
5 \times \text{AW for fluorine} & = & 5 \times 19.00 \text{ amu} = & 95.00 \text{ amu} \\
\hline
& & & 221.90 \text{ amu}
\end{array}
$$

Thus, the molar mass of IF_5 is 221.90 g. Since there are five fluorine atoms per molecule, you get

$$1.43 \text{ g } IF_5 \times \frac{1 \text{ mol } IF_5}{221.90 \text{ g } IF_5} \times \frac{6.022 \times 10^{23} \text{ } IF_5 \text{ molecules}}{1 \text{ mol } IF_5} \times \frac{5 \text{ F atoms}}{1 \text{ } IF_5 \text{ molecule}}$$

$$= 1.9\underline{4}0 \times 10^{22} = 1.94 \times 10^{22} \text{ F atoms}$$

7.69 In 100 g of the substance there is 58.8 b of barium, 13.8 g of sulfur, and 27.4 g of oxygen. Convert these quantities to moles using the molar masses as conversion factors.

$$\text{mol Ba} = 58.8 \text{ g Ba} \times \frac{1 \text{ mol Ba}}{137.33 \text{ g Ba}} = 0.4282 \text{ mol Ba}$$

$$\text{mol S} = 13.8 \text{ g S} \times \frac{1 \text{ mol S}}{32.06 \text{ g S}} = 0.4304 \text{ mol S}$$

$$\text{mol O} = 27.4 \text{ g O} \times \frac{1 \text{ mol O}}{16.00 \text{ g O}} = 1.713 \text{ mol O}$$

Divide each quantity by the smallest number of moles to get the subscripts.

$$\text{subscript for Ba} = \frac{0.4282 \text{ mol}}{0.4282 \text{ mol}} = 1.00$$

$$\text{subscript for S} = \frac{0.4304 \text{ mol}}{0.4282 \text{ mol}} = 1.01$$

$$\text{subscript for O} = \frac{1.713 \text{ mol}}{0.4282 \text{ mol}} = 4.00$$

Since all of the subscripts are whole numbers, the empirical formula is $BaSO_4$.

CHEMICAL COMPOSITION ■ 93

7.70 In 100 g of the substance there is 38.7 g of potassium, 13.8 g of nitrogen, and 47.5 g of oxygen. Convert these quantities to moles using the molar masses as conversion factors.

$$\text{mol K} = 38.7 \text{ g K} \times \frac{1 \text{ mol K}}{39.10 \text{ g K}} = 0.9898 \text{ mol K}$$

$$\text{mol N} = 13.8 \text{ g N} \times \frac{1 \text{ mol N}}{14.01 \text{ g N}} = 0.9850 \text{ mol N}$$

$$\text{mol O} = 47.5 \text{ g O} \times \frac{1 \text{ mol O}}{16.00 \text{ g O}} = 2.969 \text{ mol O}$$

Divide each quantity by the smallest number of moles to get the subscripts.

$$\text{subscript for K} = \frac{0.9898 \text{ mol}}{0.9850 \text{ mol}} = 1.01$$

$$\text{subscript for N} = \frac{0.9850 \text{ mol}}{0.9850 \text{ mol}} = 1.00$$

$$\text{subscript for O} = \frac{2.969 \text{ mol}}{0.9850 \text{ mol}} = 3.01$$

Since all of the subscripts are whole numbers, the empirical formula is KNO_3.

7.71 For potassium manganate, 100 g of the compound contains 39.7 g of potassium, 27.9 g of manganese, and 32.5 g of oxygen. Convert these quantities to moles using the molar masses as conversion factors.

$$\text{mol K} = 39.7 \text{ g K} \times \frac{1 \text{ mol K}}{39.10 \text{ g K}} = 1.015 \text{ mol K}$$

$$\text{mol Mn} = 27.9 \text{ g Mn} \times \frac{1 \text{ mol Mn}}{54.94 \text{ g Mn}} = 0.5078 \text{ mol Mn}$$

$$\text{mol O} = 32.5 \text{ g O} \times \frac{1 \text{ mol O}}{16.00 \text{ g O}} = 2.031 \text{ mol O}$$

Divide each quantity by the smallest number of moles to get the subscripts.

subscript for K = $\dfrac{1.015 \text{ mol}}{0.5078 \text{ mol}}$ = 2.00

subscript for Mn = $\dfrac{0.5078 \text{ mol}}{0.5078 \text{ mol}}$ = 1.00

subscript for O = $\dfrac{2.031 \text{ mol}}{0.5078 \text{ mol}}$ = 4.00

Since all of the subscripts are whole numbers, the empirical formula of potassium manganate is K_2MnO_4.

For potassium permanganate, 100 g of the compound contains 24.7 g of potassium, 34.8 g of manganese, and 40.5 g of oxygen. Convert these quantities to moles using the molar masses as conversion factors.

mol K = 24.7 g K × $\dfrac{1 \text{ mol K}}{39.10 \text{ g K}}$ = 0.6317 mol K

mol Mn = 34.8 g Mn × $\dfrac{1 \text{ mol Mn}}{54.94 \text{ g Mn}}$ = 0.6334 mol Mn

mol O = 40.5 g O × $\dfrac{1 \text{ mol O}}{16.00 \text{ g O}}$ = 2.531 mol O

Divide each quantity by the smallest number of moles to get the subscripts.

subscript for K = $\dfrac{0.6317 \text{ mol}}{0.6317 \text{ mol}}$ = 1.00

subscript for Mn = $\dfrac{0.6334 \text{ mol}}{0.6317 \text{ mol}}$ = 1.00

subscript for O = $\dfrac{2.531 \text{ mol}}{0.6317 \text{ mol}}$ = 4.01

Since all of the subscripts are whole numbers, the empirical formula of potassium permanganate is $KMnO_4$.

7.72 For sodium chromate, 100 g of the compound contains 28.4 g of sodium, 32.1 g of chromium, and 39.5 g of oxygen. Convert these quantities to moles using the molar masses as conversion factors.

$$\text{mol Na} = 28.4 \text{ g Na} \times \frac{1 \text{ mol Na}}{22.99 \text{ g Na}} = 1.235 \text{ mol Na}$$

$$\text{mol Cr} = 32.1 \text{ g Cr} \times \frac{1 \text{ mol Cr}}{52.00 \text{ g Cr}} = 0.6173 \text{ mol Cr}$$

$$\text{mol O} = 39.5 \text{ g O} \times \frac{1 \text{ mol O}}{16.00 \text{ g O}} = 2.469 \text{ mol O}$$

Divide each quantity by the smallest number of moles to get the subscripts.

$$\text{subscript for Na} = \frac{1.235 \text{ mol}}{0.6173 \text{ mol}} = 2.00$$

$$\text{subscript for Cr} = \frac{0.6173 \text{ mol}}{0.6173 \text{ mol}} = 1.00$$

$$\text{subscript for O} = \frac{2.469 \text{ mol}}{0.6173 \text{ mol}} = 4.00$$

Since all of the subscripts are whole numbers, the empirical formula of sodium chromate is Na_2CrO_4.

For sodium dichromate, 100 g of the compound contains 17.5 g of sodium, 39.7 g of chromium, and 42.8 g of oxygen. Convert these quantities to moles using the molar masses as conversion factors.

$$\text{mol Na} = 17.5 \text{ g Na} \times \frac{1 \text{ mol Na}}{22.99 \text{ g Na}} = 0.7612 \text{ mol Na}$$

$$\text{mol Cr} = 39.7 \text{ g Cr} \times \frac{1 \text{ mol Cr}}{52.00 \text{ g Cr}} = 0.7635 \text{ mol Cr}$$

$$\text{mol O} = 42.8 \text{ g O} \times \frac{1 \text{ mol O}}{16.00 \text{ g O}} = 2.675 \text{ mol O}$$

Divide each quantity by the smallest number of moles to get the subscripts.

subscript for Na = $\dfrac{0.7612 \text{ mol}}{0.7612 \text{ mol}}$ = 1.00

subscript for Cr = $\dfrac{0.7635 \text{ mol}}{0.7612 \text{ mol}}$ = 1.00

subscript for O = $\dfrac{2.675 \text{ mol}}{0.7612 \text{ mol}}$ = 3.51

The formula so far is $NaCrO_{3.5}$. Multiply each subscript by two to convert to whole numbers. The empirical formula of sodium dichromate is $Na_2Cr_2O_7$.

7.73 First, obtain the mass of oxygen in the compound.

mass oxygen = mass compound - mass magnesium

= 0.4145 - 0.2501 = 0.1644 g.

Thus, the sample contains 0.2501 g of magnesium and 0.1644 g of oxygen. Convert these quantities to moles using the molar masses as conversion factors.

mol Mg = 0.2501 g Mg × $\dfrac{1 \text{ mol Mg}}{24.31 \text{ g Mg}}$ = 0.01029 mol Mg

mol O = 0.1644 g O × $\dfrac{1 \text{ mol O}}{16.00 \text{ g O}}$ = 0.01028 mol O

Since both molar quantities are the same, both subscripts are equal to one, and the empirical formula is MgO.

7.74 First, obtain the mass of hydrogen in the sample.

mass hydrogen = mass compound - mass carbon = 0.3122 - 0.2880 = 0.0242 g.

Thus, the sample contains 0.2880 g of carbon and 0.0242 g of hydrogen. Convert these quantities to moles using the molar masses as conversion factors.

$$\text{mol C} = 0.2880 \text{ g C} \times \frac{1 \text{ mol C}}{12.01 \text{ g C}} = 0.02398 \text{ mol C}$$

$$\text{mol H} = 0.0242 \text{ g H} \times \frac{1 \text{ mol H}}{1.008 \text{ g H}} = 0.02401 \text{ mol H}$$

Since both molar quantities are the same, both subscripts are equal to one, and the empirical formula is CH.

7.75 First, obtain the mass of oxygen in the sample.

mass oxygen = mass compound - mass carbon = 2.200 - 0.600 = 1.600 g.

Thus, the sample contains 0.600 g of carbon and 1.600 g oxygen. Convert these quantities to moles using the molar masses as conversion factors.

$$\text{mol C} = 0.600 \text{ g C} \times \frac{1 \text{ mol C}}{12.01 \text{ g C}} = 0.04996 \text{ mol C}$$

$$\text{mol O} = 1.600 \text{ g O} \times \frac{1 \text{ mol O}}{16.00 \text{ g O}} = 0.1000 \text{ mol O}$$

Divide each quantity by the smaller number of moles to get the subscripts.

$$\text{subscript for C} = \frac{0.04996 \text{ mol}}{0.04996 \text{ mol}} = 1.00$$

$$\text{subscript for O} = \frac{0.1000 \text{ mol}}{0.04996 \text{ mol}} = 2.00$$

Since the subscripts are whole numbers, the formula of the compound is CO_2.

7.76 First, obtain the mass of oxygen in the sample.

mass oxygen = mass sample - mass hydrogen = 1.701 - 0.101 = 1.600 g.

Thus, the sample contains 1.600 g of oxygen and 0.101 g of hydrogen. Convert these quantities to moles using the molar masses as conversion factors.

$$\text{mol H} = 0.101 \text{ g H} \times \frac{1 \text{ mol H}}{1.008 \text{ g H}} = 0.1002 \text{ mol H}$$

$$\text{mol O} = 1.600 \text{ g O} \times \frac{1 \text{ mol O}}{16.00 \text{ g O}} = 0.1000 \text{ mol O}$$

Since both molar quantities are equal, both subscripts are equal to one, and the empirical formula is HO. Thus, the compound must be (HO)$_2$, or H_2O_2.

8. QUANTITIES IN CHEMICAL REACTIONS

■ Solutions to Exercises

Note on significant figures: The first time the final answer is written, it is written first with one nonsignificant figure. The least significant digit is also underlined. The final answer is then rounded to the correct number of significant figures but the least significant digit is no longer underlined.

8.1 The calculation is

$$5.82 \text{ mol } O_2 \times \frac{4 \text{ mol Fe}}{3 \text{ mol } O_2} = 7.7\underline{6}0 = 7.76 \text{ mol Fe}$$

Thus, 7.76 moles of iron are necessary to react with 5.82 moles of oxygen.

8.2 The calculation is

$$8.4 \text{ mol } C_3H_8 \times \frac{3 \text{ mol } CO_2}{1 \text{ mol } C_3H_8} = 2\underline{5}.2 = 25 \text{ mol } CO_2$$

Thus, 25 moles of carbon dioxide will be obtained when 8.4 moles of propane are burned.

8.3 The calculation is

$$2.22 \text{ mol } H_2O \times \frac{2 \text{ mol } C_8H_{18}}{18 \text{ mol } H_2O} = 0.246\underline{7} = 0.247 \text{ mol } C_8H_{18}$$

Thus, 0.247 moles of octane are required to give 2.22 moles of water.

8.4 In this exercise, besides the mole ratio, you will also need the molar masses of iron (58.55 g) and oxygen (32.00 g). The calculation is

$$3.22 \text{ g } O_2 \times \frac{1 \text{ mol } O_2}{32.00 \text{ g } O_2} \times \frac{4 \text{ mol Fe}}{3 \text{ mol } O_2} \times \frac{55.85 \text{ g Fe}}{1 \text{ mol Fe}} = 7.493 = 7.49 \text{ g Fe}$$

Therefore, 7.49 g of iron are required to react with 3.22 g of oxygen.

8.5 In this exercise, besides the mole ratio, you will also need the molar masses of phosphorus (123.88 g) and phosphorus pentachloride (208.22 g). The calculation is

$$5.00 \text{ g } P_4 \times \frac{1 \text{ mol } P_4}{123.88 \text{ g } P_4} \times \frac{4 \text{ mol } PCl_5}{1 \text{ mol } P_4} \times \frac{208.22 \text{ g } PCl_5}{1 \text{ mol } PCl_5} = 33.62 = 33.6 \text{ g } PCl_5$$

Thus, 33.6 g of PCl_5 can be produced from 5.00 g of P_4.

8.6 In this exercise, besides the mole ratio, you will need the molar masses of oxygen (32.00 g) and nitrogen dioxide (46.01 g). The calculation is

$$141 \text{ g } NO_2 \times \frac{1 \text{ mol } NO_2}{46.01 \text{ g } NO_2} \times \frac{1 \text{ mol } O_2}{2 \text{ mol } NO_2} \times \frac{32.00 \text{ g } O_2}{1 \text{ mol } O_2} = 49.03 = 49.0 \text{ g } O_2$$

Thus, 49.0 g of oxygen are used up when 141 g of nitrogen dioxide are formed.

8.7 a. Calculate the amount of H_2 that is required to react with 27 moles of N_2

$$27 \text{ mol } N_2 \times \frac{3 \text{ mol } H_2}{1 \text{ mol } N_2} = 81 \text{ mol } H_2$$

The available quantity of H_2 (81 mol) is exactly the right amount, so there is no limiting reactant.

b. The amount of H_2 that is required to react with 2.5 moles of N_2 is

$$2.5 \text{ mol } N_2 \times \frac{3 \text{ mol } H_2}{1 \text{ mol } N_2} = 7.5 \text{ mol } H_2$$

The available quantity of H_2 (6.2 mol) is less than the required amount, so H_2 is the limiting reactant.

c. The amount of H_2 that is required to react with 1.7 moles of N_2 is

$$1.7 \text{ mol } N_2 \times \frac{3 \text{ mol } H_2}{1 \text{ mol } N_2} = 5.1 \text{ mol } H_2$$

The available quantity of H_2 (9.00 mol) is more than the required amount, so N_2 is the limiting reactant.

8.8 First, determine the limiting reactant, if there is one. Find the amount of HCl that is required to react with 1.0 mole of $CaCO_3$.

$$1.0 \text{ mol } CaCO_3 \times \frac{2 \text{ mol HCl}}{1 \text{ mol } CaCO_3} = 2.0 \text{ mol HCl}$$

Since the amount of HCl available (1.0 mol) is less than the required amount, HCl is the limiting reactant. Use the available quantity of HCl to calculate the quantity of each product formed.

$$1.0 \text{ mol HCl} \times \frac{1 \text{ mol } CaCl_2}{2 \text{ mol HCl}} = 0.50 \text{ mol } CaCl_2$$

$$1.0 \text{ mol HCl} \times \frac{1 \text{ mol } CO_2}{2 \text{ mol HCl}} = 0.50 \text{ mol } CO_2$$

$$1.0 \text{ mol HCl} \times \frac{1 \text{ mol } H_2O}{2 \text{ mol HCl}} = 0.50 \text{ mol } H_2O$$

Next, calculate the quantity of $CaCO_3$ that is used up in the reaction.

$$1.0 \text{ mol HCl} \times \frac{1 \text{ mol } CaCO_3}{2 \text{ mol HCl}} = 0.50 \text{ mol } CaCO_3$$

The quantity of $CaCO_3$ remaining after the reaction is

1.0 mol $CaCO_3$ - 0.50 mol $CaCO_3$ = 0.50 = 0.5 mol $CaCO_3$ (remaining)

102 ■ CHAPTER 8

8.9 First, convert the mass of each reactant to moles, using the molar masses as conversion factors. The molar masses of $NaHCO_3$ and HCl are 84.01 g and 36.46 g, respectively. Therefore,

$$1.00 \text{ g NaHCO}_3 \times \frac{1 \text{ mol NaHCO}_3}{84.01 \text{ g NaHCO}_3} = 0.01190 \text{ mol NaHCO}_3$$

$$2.00 \text{ g HCl} \times \frac{1 \text{ mol HCl}}{36.46 \text{ g HCl}} = 0.05485 \text{ mol HCl}$$

Next, calculate the required amount of HCl necessary to react with the 0.01190 mole of $NaHCO_3$.

$$0.01190 \text{ mol NaHCO}_3 \times \frac{1 \text{ mol HCl}}{1 \text{ mol NaHCO}_3} = 0.01190 \text{ mol HCl}$$

Since the available quantity of HCl is more than the required amount, $NaHCO_3$ is the limiting reactant. Using this quantity, calculate the mass of carbon dioxide that is produced. The molar mass of CO_2 is 44.01 g. Thus,

$$0.01190 \text{ mol NaHCO}_3 \times \frac{1 \text{ mol CO}_2}{1 \text{ mol NaHCO}_3} \times \frac{44.01 \text{ g CO}_2}{1 \text{ mol CO}_2} = 0.5237 = 0.524 \text{ g CO}_2$$

8.10 The reaction is

$$2 H_2(g) + O_2(g) \longrightarrow 2 H_2O(g)$$

First, convert the mass of hydrogen to moles using the molar mass of H_2 (2.016 g).

$$5.00 \text{ g H}_2 \times \frac{1 \text{ mol H}_2}{2.016 \text{ g H}_2} = 2.480 \text{ mol H}_2$$

Next, calculate the theoretical yield of H_2O. The molar mass of H_2O is 18.02 g. Thus,

$$2.480 \text{ mol H}_2 \times \frac{2 \text{ mol H}_2O}{2 \text{ mol H}_2} \times \frac{18.02 \text{ g H}_2O}{1 \text{ mol H}_2O} = 44.69 \text{ g H}_2O$$

Therefore, the percentage yield is

$$\text{percentage yield of } H_2O = \frac{0.32 \text{ g}}{44.69 \text{ g}} \times 100\% = 0.72\%$$

QUANTITIES IN CHEMICAL REACTIONS ■ 103

■ Answers to Questions to Test Your Reading

8.1 Dalton's atomic theory states that atoms in molecules are neither manufactured nor lost during a chemical reaction; they are merely rearranged. In a balanced equation, every atom on the left side of the equation also appears on the right side.

8.2 For the reaction

$$CO(g) + 3\ H_2(g) \longrightarrow CH_4(g) + H_2O(g)$$

In terms of molecules, one molecule of carbon monoxide (CO) reacts with three molecules of hydrogen (H_2) to form one molecule of methane (CH_4) and one molecule of water (H_2O).

In terms of moles, one mole of carbon monoxide reacts with three moles of hydrogen to form one mole of methane and two moles of water.

8.3 The chemical equation for this reaction is

$$C_2H_4(g) + 3\ O_2(g) \longrightarrow 2\ CO_2(g) + 2\ H_2O(g)$$

In terms of molecules, one molecule of ethylene (C_2H_4) reacts with three molecules of oxygen (O_2) to form two molecules of carbon dioxide (CO_2) and two molecules of water (H_2O).

In terms of moles, one mole of ethylene reacts with three moles of oxygen to form two moles of carbon dioxide and two moles of water.

8.4 The chemical equation for the reaction is

$$2\ H_2S(g) + 3\ O_2(g) \longrightarrow 2\ SO_2(g) + 2\ H_2O(g)$$

The number of moles of H_2S that will combine with 15 moles of oxygen is

$$15\ \text{mol}\ O_2 \times \frac{2\ \text{mol}\ H_2S}{3\ \text{mol}\ O_2} = 10.\ \text{mol}\ H_2S$$

Thus, the statement that 5.0 mol of H_2S will combine with 15 mol of O_2 is false.

8.5 A mole ratio is a conversion factor that relates the number of moles of one of the reactants or products to the number of moles of any other reactant or product. The coefficients of the balanced chemical equation appear in the mole ratio.

8.6 Since mole ratios are conversion factors, the desired unit must be on the top and the given unit on the bottom. In this question, the given unit is moles of hydrogen and the desired unit is moles of carbon monoxide. Thus, the correct mole ratio must be

$$\frac{1 \text{ mol CO}}{3 \text{ mol H}_2}$$

8.7 In each of these reactions, the given unit (reactant) is moles of oxygen and the desired unit (product) is carbon dioxide. Thus, each mole ratio will have mol CO_2 on top and mol O_2 on the bottom.

a. $2 \text{ CO} + O_2 \longrightarrow 2 \text{ CO}_2$ The mole ratio is: $\dfrac{2 \text{ mol CO}_2}{1 \text{ mol O}_2}$

b. $CH_4 + 2 O_2 \longrightarrow CO_2 + 2 H_2O$ The mole ratio is: $\dfrac{1 \text{ mol CO}_2}{2 \text{ mol O}_2}$

c. $2 C_2H_6 + 7 O_2 \longrightarrow 4 CO_2 + 6 H_2O$ The mole ratio is: $\dfrac{4 \text{ mol CO}_2}{7 \text{ mol O}_2}$

d. $C_3H_8 + 5 O_2 \longrightarrow 3 CO_2 + 4 H_2O$ The mole ratio is: $\dfrac{3 \text{ mol CO}_2}{5 \text{ mol O}_2}$

8.8 In these reactions, moles of product will be on top of the mole ratio and moles of reactant will be on the bottom.

a. $2 SO_2 + O_2 \longrightarrow 2 SO_3$ The mole ratio is: $\dfrac{2 \text{ mol SO}_3}{2 \text{ mol SO}_2}$

b. $CS_2 + 3 Cl_2 \longrightarrow CCl_4 + S_2Cl_2$ The mole ratio is: $\dfrac{1 \text{ mol CCl}_4}{3 \text{ mol Cl}_2}$

c. $WO_3 + 3 H_2 \longrightarrow W + 3 H_2O$ The mole ratio is: $\dfrac{3 \text{ mol H}_2O}{1 \text{ mol WO}_3}$

d. $4 Li + O_2 \longrightarrow 2 Li_2O$ The mole ratio is: $\dfrac{2 \text{ mol Li}_2O}{4 \text{ mol Li}}$

QUANTITIES IN CHEMICAL REACTIONS 105

8.9 The reaction in question 8.2 is

$$CO(g) + 3\ H_2(g) \longrightarrow CH_4(g) + H_2O(g)$$

According to this reaction, 1 <u>mole</u> of carbon monoxide reacts with 3 <u>moles</u> of hydrogen. Coefficients in chemical reactions never refer to masses. In order to determine the mass relationships, you must use the molar masses of the substances in the calculation.

8.10 A <u>limiting reactant</u> is the reactant that is used up when a reaction goes to completion even though other reactants are not consumed completely.

8.11 The main difference is that when stoichiometric quantities are used, every reactant is used up when the reaction is completed, whereas when a limiting reactant is used, it is used up and the other reactants will be left over when the reaction is completed.

8.12 For a given reaction, the <u>theoretical yield</u> is the maximum mass of a product that can be obtained from a given amounts of reactants. The <u>actual yield</u> is the mass of product that is actually recovered in an experiment. The <u>percentage yield</u> is the actual yield expressed as a percentage of the theoretical yield, or

$$\text{percentage yield} = \frac{\text{actual yield}}{\text{theoretical yield}} \times 100\%$$

■ Solutions to Practice Problems

<u>Note on significant figures:</u> The first time the final answer is written, it is written first with one nonsignificant figure. The least significant digit is also underlined. The final answer is then rounded to the correct number of significant figures but the least significant digit is no longer underlined.

8.13 $1.86\ \text{mol Al} \times \dfrac{3\ \text{mol}\ O_2}{4\ \text{mol Al}} = 1.3\underline{9}5 = 1.40\ \text{mol}\ O_2$

8.14 $2.54\ \text{mol}\ H_2 \times \dfrac{1\ \text{mol}\ N_2}{3\ \text{mol}\ H_2} = 0.84\underline{6}7 = 0.847\ \text{mol}\ N_2$

8.15 First, balance the equation.

$$C_2H_5OH(g) + 3\ O_2(g) \longrightarrow 2\ CO_2(g) + 3\ H_2O(g)$$

Thus, the moles of oxygen required to react with 8.24 mol ethanol is

$$8.24\ \text{mol}\ C_2H_5OH \times \frac{3\ \text{mol}\ O_2}{1\ \text{mol}\ C_2H_5OH} = 24.\underline{7}2 = 24.7\ \text{mol}\ O_2$$

8.16 First, balance the equation.

$$2\ C_8H_{18}(g) + 25\ O_2(g) \longrightarrow 16\ CO_2(g) + 18\ H_2O(g)$$

Thus, the moles of octane required to react with 4.44 mol oxygen is

$$4.44\ \text{mol}\ O_2 \times \frac{2\ \text{mol}\ C_8H_{18}}{25\ \text{mol}\ O_2} = 0.355\underline{2} = 0.355\ \text{mol}\ C_8H_{18}$$

8.17 $5.6\ \text{mol}\ I_2 \times \dfrac{2\ \text{mol}\ AlI_3}{3\ \text{mol}\ I_2} = 3.\underline{7}3 = 3.7\ \text{mol}\ AlI_3$

8.18 $24.3\ \text{mol}\ H_2SO_4 \times \dfrac{1\ \text{mol}\ Al_2(SO_4)_3}{3\ \text{mol}\ H_2SO_4} = 8.1\underline{0}0 = 8.10\ \text{mol}\ Al_2(SO_4)_3$

8.19 $3.00\ \text{mol}\ NaOH \times \dfrac{1\ \text{mol}\ Na_2SO_4}{2\ \text{mol}\ NaOH} = 1.5\underline{0}0 = 1.50\ \text{mol}\ Na_2SO_4$

8.20 $5.22\ \text{mol}\ Al \times \dfrac{2\ \text{mol}\ Al_2O_3}{4\ \text{mol}\ Al} = 2.6\underline{1}0 = 2.61\ \text{mol}\ Al_2O_3$

8.21 $3.56\ \text{mol}\ HNO_3 \times \dfrac{3\ \text{mol}\ NO_2}{2\ \text{mol}\ HNO_3} = 5.3\underline{4}0 = 5.34\ \text{mol}\ NO_2$

8.22 $134\ \text{mol}\ P_4 \times \dfrac{10\ \text{mol}\ C}{1\ \text{mol}\ P_4} = 1.3\underline{4}0 \times 10^3 = 1.34 \times 10^3\ \text{mol}\ C$

8.23 The moles of sodium carbonate required are

$$1.60 \text{ mol NaHCO}_3 \times \frac{1 \text{ mol Na}_2\text{CO}_3}{2 \text{ mol NaHCO}_3} = 0.8000 = 0.800 \text{ mol Na}_2\text{CO}_3$$

For carbon dioxide, the required amount is

$$1.60 \text{ mol NaHCO}_3 \times \frac{1 \text{ mol CO}_2}{2 \text{ mol NaHCO}_3} = 0.8000 = 0.800 \text{ mol CO}_2$$

8.24 The moles of cesium sulfate required are

$$1.76 \text{ mol CsOH} \times \frac{1 \text{ mol Cs}_2\text{SO}_4}{2 \text{ mol CsOH}} = 0.8800 = 0.880 \text{ mol Cs}_2\text{SO}_4$$

For barium hydroxide, the required amount is

$$1.76 \text{ mol CsOH} \times \frac{1 \text{ mol Ba(OH)}_2}{2 \text{ mol CsOH}} = 0.8800 = 0.880 \text{ mol Ba(OH)}_2$$

8.25 The molar masses of sodium (Na) and chlorine (Cl_2) are 22.99 g and 70.90 g, respectively. Thus,

$$1.00 \text{ g Cl}_2 \times \frac{1 \text{ mol Cl}_2}{70.90 \text{ g Cl}_2} \times \frac{2 \text{ mol Na}}{1 \text{ mol Cl}_2} \times \frac{22.99 \text{ g Na}}{1 \text{ mol Na}} = 0.6485 = 0.649 \text{ g Na}$$

8.26 The molar masses of iron (Fe) and oxygen (O_2) are 55.85 g and 32.00 g, respectively. Thus,

$$1.00 \text{ g Fe} \times \frac{1 \text{ mol Fe}}{55.85 \text{ g Fe}} \times \frac{3 \text{ mol O}_2}{4 \text{ mol Fe}} \times \frac{32.00 \text{ g O}_2}{1 \text{ mol O}_2} = 0.4297 = 0.430 \text{ g O}_2$$

8.27 The molar masses of chromium (Cr) and hydrochloric acid (HCl) are 52.00 g and 36.46 g, respectively. Thus,

$$1.00 \text{ g Cr} \times \frac{1 \text{ mol Cr}}{52.00 \text{ g Cr}} \times \frac{2 \text{ mol HCl}}{1 \text{ mol Cr}} \times \frac{36.46 \text{ g HCl}}{1 \text{ mol HCl}} = 1.402 = 1.40 \text{ g HCl}$$

8.28 The molar masses of magnesium (Mg) and oxygen (O_2) are 24.31 g and 32.00 g, respectively. Thus,

$$1.00 \text{ g Mg} \times \frac{1 \text{ mol Mg}}{24.31 \text{ g Mg}} \times \frac{1 \text{ mol O}_2}{2 \text{ mol Mg}} \times \frac{32.00 \text{ g O}_2}{1 \text{ mol O}_2} = 0.6582 = 0.658 \text{ g O}_2$$

8.29 The molar masses of calcium carbonate ($CaCO_3$) and calcium oxide (CaO) are 100.09 g and 56.08 g, respectively. Thus,

$$1.62 \text{ g } CaCO_3 \times \frac{1 \text{ mol } CaCO_3}{100.09 \text{ g } CaCO_3} \times \frac{1 \text{ mol } CaO}{1 \text{ mol } CaCO_3} \times \frac{56.08 \text{ g } CaO}{1 \text{ mol } CaO}$$

$$= 0.9077 = 0.908 \text{ g } CaO$$

8.30 The molar masses of copper(II) oxide (CuO) and oxygen (O_2) are 79.55 g and 32.00 g, respectively. Thus,

$$2.64 \text{ g } CuO \times \frac{1 \text{ mol } CuO}{79.55 \text{ g } CuO} \times \frac{1 \text{ mol } O_2}{4 \text{ mol } CuO} \times \frac{32.00 \text{ g } O_2}{1 \text{ mol } O_2} = 0.2654 = 0.265 \text{ g } O_2$$

8.31 The molar masses of mercury(II) oxide (HgO) and mercury (Hg) are 216.59 g and 200.59 g, respectively. Thus,

$$10.3 \text{ g } HgO \times \frac{1 \text{ mol } HgO}{216.59 \text{ g } HgO} \times \frac{2 \text{ mol } Hg}{2 \text{ mol } HgO} \times \frac{200.59 \text{ g } Hg}{1 \text{ mol } Hg} = 9.539 = 9.54 \text{ g } Hg$$

8.32 The molar masses of water (H_2O) and oxygen (O_2) are 18.02 g and 32.00 g, respectively. Thus,

$$1.32 \text{ g } H_2O \times \frac{1 \text{ mol } H_2O}{18.02 \text{ g } H_2O} \times \frac{1 \text{ mol } O_2}{2 \text{ mol } H_2O} \times \frac{32.00 \text{ g } O_2}{1 \text{ mol } O_2} = 1.172 = 1.17 \text{ g } O_2$$

8.33 The molar masses of hydrochloric acid (HCl) and ammonium chloride (NH_4Cl) are 36.46 g and 53.49 g, respectively. Thus,

$$2.36 \text{ g } NH_4Cl \times \frac{1 \text{ mol } NH_4Cl}{53.49 \text{ g } NH_4Cl} \times \frac{1 \text{ mol } HCl}{1 \text{ mol } NH_4Cl} \times \frac{36.46 \text{ g } HCl}{1 \text{ mol } HCl}$$

$$= 1.609 = 1.61 \text{ g } HCl$$

8.34 The molar masses of sodium chloride ($NaCl$) and sodium hydroxide ($NaOH$) are 58.44 g and 40.00 g, respectively. Thus,

$$23.1 \text{ g } NaCl \times \frac{1 \text{ mol } NaCl}{58.44 \text{ g } NaCl} \times \frac{1 \text{ mol } NaOH}{1 \text{ mol } NaCl} \times \frac{40.00 \text{ g } NaOH}{1 \text{ mol } NaOH}$$

$$= 15.81 = 15.8 \text{ g } NaOH$$

8.35 The molar masses of sodium (Na) and hydrogen (H_2) are 22.99 g and 2.016 g, respectively. Thus,

$$3.5 \text{ g } H_2 \times \frac{1 \text{ mol } H_2}{2.016 \text{ g } H_2} \times \frac{2 \text{ mol Na}}{1 \text{ mol } H_2} \times \frac{22.99 \text{ g Na}}{1 \text{ mol Na}} = 79.8 = 80. \text{ g Na}$$

8.36 The molar masses of magnesium hydroxide ($Mg(OH)_2$) and water (H_2O) are 58.33 g and 18.02 g, respectively. Thus,

$$13.8 \text{ g } Mg(OH)_2 \times \frac{1 \text{ mol } Mg(OH)_2}{58.33 \text{ g } Mg(OH)_2} \times \frac{2 \text{ mol } H_2O}{1 \text{ mol } Mg(OH)_2} \times \frac{18.02 \text{ g } H_2O}{1 \text{ mol } H_2O}$$

$$= 8.526 = 8.53 \text{ g } H_2O$$

8.37 In each part, calculate the required amount of water necessary to react with the given amount of KO_2.

a. $6.4 \text{ mol } KO_2 \times \dfrac{2 \text{ mol } H_2O}{4 \text{ mol } KO_2} = 3.2 \text{ mol } H_2O$

Since the available amount of water (2.1 mol) is less than the required amount, H_2O is the limiting reactant.

b. $8.4 \text{ mol } KO_2 \times \dfrac{2 \text{ mol } H_2O}{4 \text{ mol } KO_2} = 4.2 \text{ mol } H_2O$

Since the available amount of water (1.5 mol) is less than the required amount, H_2O is the limiting reactant.

c. $8.4 \text{ mol } KO_2 \times \dfrac{2 \text{ mol } H_2O}{4 \text{ mol } KO_2} = 4.2 \text{ mol } H_2O$

Since the available amount of water (2.1 mol) is less than the required amount, H_2O is the limiting reactant.

8.38 In each part, calculate the required amount of oxygen necessary to react with the given amount of ammonia.

a. $4.8 \text{ mol } NH_3 \times \dfrac{5 \text{ mol } O_2}{4 \text{ mol } NH_3} = 6.0 \text{ mol } O_2$

Since the given amount of oxygen (5.8 mol) is less than the required amount, O_2 is the limiting reactant.

b. 12 mol NH$_3$ × $\dfrac{5 \text{ mol O}_2}{4 \text{ mol NH}_3}$ = 15 mol O$_2$

Since the given amount of oxygen (15 mol) is exactly the right amount, there is no limiting reactant.

c. 32 mol NH$_3$ × $\dfrac{5 \text{ mol O}_2}{4 \text{ mol NH}_3}$ = 40. mol O$_2$

Since the given amount of oxygen (44 mol) is more than the required amount, NH$_3$ is the limiting reactant.

8.39 a. The balanced equation is

$$P_4(s) + 6 \text{ Cl}_2(g) \longrightarrow 4 \text{ PCl}_3(s)$$

The amount of Cl$_2$ required to react with 2.00 mol P$_4$ is

2.0 mol P$_4$ × $\dfrac{6 \text{ mol Cl}_2}{1 \text{ mol P}_4}$ = 12 mol Cl$_2$

Since the given amount of Cl$_2$ (2.00 mol) is less than the required amount, Cl$_2$ is the limiting reactant.

b. The balanced equation is

$$2 \text{ Al}(s) + 3 \text{ Cl}_2(g) \longrightarrow 2 \text{ AlCl}_3(s)$$

The amount of Cl$_2$ required to react with 2.0 mol Al is

2.0 mol Al × $\dfrac{3 \text{ mol Cl}_2}{2 \text{ mol Al}}$ = 3.0 mol Cl$_2$

Since the given amount of Cl$_2$ (2.0 mol) is less than the required amount, Cl$_2$ is the limiting reactant.

c. The balanced equation is

$$C(s) + 2\ Cl_2(g) \longrightarrow CCl_4(l)$$

The amount of Cl_2 required to react with 2.0 mol C is

$$2.0\ \text{mol C} \times \frac{2\ \text{mol } Cl_2}{1\ \text{mol C}} = 4.0\ \text{mol } Cl_2$$

Since the given amount of Cl_2 (2.0 mol) is less than the required amount, Cl_2 is the limiting reactant.

8.40 a. The balanced equation is

$$2\ C(s) + O_2(g) \longrightarrow 2\ CO(g)$$

The amount of O_2 required to react with 2.0 mol C is

$$2.0\ \text{mol C} \times \frac{1\ \text{mol } O_2}{2\ \text{mol C}} = 1.0\ \text{mol } O_2$$

Since the given amount of O_2 (2.0 mol) is more than the required amount, C is the limiting reactant.

b. The balanced equation is

$$2\ Fe(s) + 3\ Cl_2(g) \longrightarrow 2\ FeCl_3(s)$$

The amount of Cl_2 required to react with 2.0 mol Fe is

$$2.0\ \text{mol Fe} \times \frac{3\ \text{mol } Cl_2}{2\ \text{mol Fe}} = 3.0\ \text{mol } Cl_2$$

Since the given amount of Cl_2 (2.0 mol) is less than the required amount, Cl_2 is the limiting reactant.

c. The balanced equation is

$$N_2(g) + 3\ H_2(g) \longrightarrow 2\ NH_3(g)$$

The amount of H_2 required to react with 2.0 mol N_2 is

$$2.0\ \text{mol } N_2 \times \frac{3\ \text{mol } H_2}{1\ \text{mol } N_2} = 6.0\ \text{mol } H_2$$

Since the given amount of H_2 (2.0 mol) is less than the required amount, H_2 is the limiting reactant.

8.41 First, determine the limiting reactant, if there is one. Calculate the required amount of NH_3 necessary to react with 3.0 mol H_2SO_4.

$$3.0 \text{ mol Na} \times \frac{1 \text{ mol Cl}_2}{2 \text{ mol Na}} = 1.5 \text{ mol Cl}_2$$

Since the given amount of Cl_2 (3.0 mol) is more than the required amount, Na is the limiting reactant, and will be used up in the reaction. Next, calculate the amount of NaCl that will be formed.

$$3.0 \text{ mol Na} \times \frac{2 \text{ mol NaCl}}{2 \text{ mol Na}} = 3.0 \text{ mol NaCl}$$

Finally, determine the amount of Cl_2 that will remain after the reaction is over. The amount used up in the reaction is 1.5 mol. Thus,

$$3.0 \text{ mol Cl}_2 - 1.5 \text{ mol Cl}_2 = 1.5 \text{ mol Cl}_2 \qquad \text{(remaining)}$$

8.42 First, determine the limiting reactant, if there is one. Calculate the number of moles of O_2 necessary to react with 1.00 mol Fe.

$$1.00 \text{ mol Fe} \times \frac{3 \text{ mol O}_2}{4 \text{ mol Fe}} = 0.750 \text{ mol O}_2$$

Since the moles of O_2 available (1.00 mol) is more than the required amount, Fe is the limiting reactant, and will be used up in the reaction. Next, use the limiting reactant to calculate the amount of Fe_2O_3 that is formed.

$$1.00 \text{ mol Fe} \times \frac{2 \text{ mol Fe}_2O_3}{4 \text{ mol Fe}} = 0.500 \text{ mol Fe}_2O_3$$

Finally, determine the amount of O_2 that will remain after the reaction is over. The amount used up in the reaction is 0.750 mol. Thus,

$$1.00 \text{ mol O}_2 - 0.750 \text{ mol O}_2 = 0.2\underline{5}0 = 0.25 \text{ mol O}_2 \qquad \text{(remaining)}$$

8.43 First, determine the limiting reactant, if there is one. Calculate the number of moles of Al required to react with 2.5 mol O_2.

$$2.5 \text{ mol } O_2 \times \frac{4 \text{ mol Al}}{3 \text{ mol } O_2} = 3.33 \text{ mol Al}$$

Since the available amount of Al (2.5 mol) is less than the required amount, Al is the limiting reactant, and is used up in the reaction. Next, use the limiting reactant to calculate the amount of Al_2O_3 that is formed.

$$2.5 \text{ mol Al} \times \frac{2 \text{ mol } Al_2O_3}{4 \text{ mol Al}} = 1.25 = 1.3 \text{ mol } Al_2O_3$$

Finally, determine the amount of O_2 that will remain after the reaction is over. The amount used up in the reaction is

$$2.5 \text{ mol Al} \times \frac{3 \text{ mol } O_2}{4 \text{ mol Al}} = 1.88 \text{ mol } O_2$$

Thus, the amount remaining is

2.5 mol O_2 - 1.88 mol O_2 = 0.62 mol O_2 (remaining)

8.44 First, determine the limiting reactant, if there is one. Calculate the amount of O_2 required to react with 3.00 mol Cu_2O.

$$3.00 \text{ mol } Cu_2O \times \frac{1 \text{ mol } O_2}{2 \text{ mol } Cu_2O} = 1.50 \text{ mol } O_2$$

Since the available amount of O_2 (3.00 mol) is more than the required amount, Cu_2O is the limiting reactant, and is used up in the reaction. Next, use the limiting reactant to calculate the amount of CuO that will be formed.

$$3.00 \text{ mol } Cu_2O \times \frac{4 \text{ mol CuO}}{2 \text{ mol } Cu_2O} = 6.00 \text{ mol CuO}$$

Finally, determine the amount of O_2 that will remain after the reaction is over. The amount used up in the reaction is

$$3.00 \text{ mol } Cu_2O \times \frac{1 \text{ mol } O_2}{2 \text{ mol } Cu_2O} = 1.50 \text{ mol } O_2$$

Thus, the amount remaining is

3.00 mol O_2 - 1.50 mol O_2 = 1.50 mol O_2 (remaining)

8.45 First, convert the given mass to moles. The molar masses of NH_3 and HCl are 17.03 g and 36.46 g, respectively. Thus,

$$1.00 \text{ g } NH_3 \times \frac{1 \text{ mol } NH_3}{17.03 \text{ g } NH_3} = 0.05872 \text{ mol } NH_3$$

$$1.00 \text{ g HCl} \times \frac{1 \text{ mol HCl}}{36.46 \text{ g HCl}} = 0.02743 \text{ mol HCl}$$

Next, determine the limiting reactant, if there is one. Calculate the amount of HCl required to react with 0.05872 mol NH_3.

$$0.05872 \text{ mol } NH_3 \times \frac{1 \text{ mol HCl}}{1 \text{ mol } NH_3} = 0.05872 \text{ mol HCl}$$

Since the available amount of HCl (0.02743 mol) is less than the required amount, HCl is the limiting reactant. Use the limiting reactant to determine the amount of NH_3 that will be used up in the reaction.

$$0.02743 \text{ mol HCl} \times \frac{1 \text{ mol } NH_3}{1 \text{ mol HCl}} \times \frac{17.03 \text{ g } NH_3}{1 \text{ mol } NH_3} = 0.4671 \text{ g } NH_3$$

Thus, the amount of NH_3 present at the end of the reaction is

$$1.00 \text{ g } NH_3 - 0.467 \text{ g } NH_3 = 0.533 = 0.53 \text{ g } NH_3 \qquad \text{(remaining)}$$

8.46 First, convert the given masses of reactants to moles. The molar masses of NaOH and HCl are 40.00 g and 36.46 g, respectively. Thus

$$1.00 \text{ g NaOH} \times \frac{1 \text{ mol NaOH}}{40.00 \text{ g NaOH}} = 0.02500 \text{ mol NaOH}$$

$$1.00 \text{ g HCl} \times \frac{1 \text{ mol HCl}}{36.46 \text{ g HCl}} = 0.02743 \text{ mol HCl}$$

Next, determine the limiting reactant, if there is one. Calculate the amount of HCl required to react with 0.02500 mol NaOH.

$$0.02500 \text{ mol NaOH} \times \frac{1 \text{ mol HCl}}{1 \text{ mol NaOH}} = 0.02500 \text{ mol NaOH}$$

Since the available amount of HCl (0.02743 mol) is more than the required amount, NaOH is the limiting reactant. Use the limiting reactant to calculate the amount of NaCl that will form when the reaction is complete. The molar mass of NaCl is 58.44 g. Thus,

$$0.02500 \text{ mol NaOH} \times \frac{1 \text{ mol NaCl}}{1 \text{ mol NaOH}} \times \frac{58.44 \text{ g NaCl}}{1 \text{ mol NaCl}} = 1.461 = 1.46 \text{ g NaCl}$$

QUANTITIES IN CHEMICAL REACTIONS ■ 115

8.47 First, convert the given masses to moles. The molar masses of CO and H_2 are 28.01 g and 2.016 g, respectively. Thus,

$$30.0 \text{ g CO} \times \frac{1 \text{ mol CO}}{28.01 \text{ g CO}} = 1.071 \text{ mol CO}$$

$$30.0 \text{ g } H_2 \times \frac{1 \text{ mol } H_2}{2.016 \text{ g } H_2} = 14.88 \text{ mol } H_2$$

Next, determine the limiting reactant, if there is one. Calculate the amount of H_2 required to react with 1.071 mol CO.

$$1.071 \text{ mol CO} \times \frac{2 \text{ mol } H_2}{1 \text{ mol CO}} = 2.142 \text{ mol } H_2$$

Since the available amount of H_2 (14.88 mol) is more than the required amount, CO is the limiting reactant. Next, use the limiting reactant to calculate the mass of CH_3OH that will be formed and the mass of H_2 that will react. The molar mass of CH_3OH is 32.04 g.

$$1.071 \text{ mol CO} \times \frac{1 \text{ mol } CH_3OH}{1 \text{ mol CO}} \times \frac{32.04 \text{ g } CH_3OH}{1 \text{ mol } CH_3OH} = 34.31 = 34.3 \text{ g } CH_3OH$$

$$1.071 \text{ mol CO} \times \frac{2 \text{ mol } H_2}{1 \text{ mol CO}} \times \frac{2.016 \text{ g } H_2}{1 \text{ mol } H_2} = 4.318 \text{ g } H_2$$

Finally, determine the mass of H_2 remaining when the reaction is over.

$$30.0 \text{ g } H_2 - 4.318 \text{ g } H_2 = 25.\underline{6}8 = 25.7 \text{ g } H_2 \qquad \text{(remaining)}$$

8.48 First, convert the given masses to moles. The molar masses of CS_2 and O_2 are 76.13 g and 32.00 g, respectively. Thus,

$$254 \text{ g } CS_2 \times \frac{1 \text{ mol } CS_2}{76.13 \text{ g } CS_2} = 3.336 \text{ mol } CS_2$$

$$254 \text{ g } O_2 \times \frac{1 \text{ mol } O_2}{32.00 \text{ g } O_2} = 7.938 \text{ mol } O_2$$

Next, determine the limiting reactant, if there is one. Calculate the amount of O_2 required to react with 3.336 mol CS_2.

$$3.336 \text{ mol } CS_2 \times \frac{3 \text{ mol } O_2}{1 \text{ mol } CS_2} = 10.01 \text{ mol } O_2$$

Since the available amount of O_2 (7.938 mol) is less than the required amount, O_2 is the limiting reactant.

116 ■ CHAPTER 8

Next, use the limiting reactant to calculate the mass of CS_2 that will react and the mass of CO_2 that will be formed. The molar mass of CO_2 is 44.01 g. Thus,

$$7.938 \text{ mol } O_2 \times \frac{1 \text{ mol } CS_2}{3 \text{ mol } O_2} \times \frac{76.13 \text{ g } CS_2}{1 \text{ mol } CS_2} = 201.4 \text{ g } CS_2$$

$$7.938 \text{ mol } O_2 \times \frac{1 \text{ mol } CO_2}{3 \text{ mol } O_2} \times \frac{44.01 \text{ g } CO_2}{1 \text{ mol } CO_2} = 116.4 = 116 \text{ g } CO_2$$

Finally, determine the mass of CS_2 remaining when the reaction is over.

$$254 \text{ g } CS_2 - 201.4 \text{ g } CS_2 = 52.6 = 53 \text{ g } CS_2 \qquad \text{(remaining)}$$

8.49 percentage yield $= \dfrac{4.72 \text{ g}}{4.87 \text{ g}} \times 100\% = 96.92 = 96.9\%$

8.50 percentage yield $= \dfrac{4.02 \text{ g}}{112 \text{ g}} \times 100\% = 3.589 = 3.59\%$

8.51 First, calculate the theoretical yield. The molar mass of nitrogen dioxide, NO_2, is 46.01 g, and for HNO_3 it is 63.02 g.

$$60.0 \text{ g } NO_2 \times \frac{1 \text{ mol } NO_2}{46.01 \text{ g } NO_2} \times \frac{2 \text{ mol } HNO_3}{3 \text{ mol } NO_2} \times \frac{63.02 \text{ g } HNO_3}{1 \text{ mol } HNO_3}$$

$$= 54.79 = 54.8 \text{ g } HNO_3$$

Thus, the percentage yield of HNO_3 is

percentage yield $= \dfrac{44.2 \text{ g}}{54.79 \text{ g}} \times 100\% = 80.67 = 80.7\%$

8.52 First, calculate the theoretical yield of nitrobenzene ($C_6H_5NO_2$). The molar mass is 123.11 g, and for benzene (C_6H_6) it is 78.11 g.

$$20.3 \text{ g } C_6H_6 \times \frac{1 \text{ mol } C_6H_6}{78.11 \text{ g } C_6H_6} \times \frac{1 \text{ mol } C_6H_5NO_2}{1 \text{ mol } C_6H_6} \times \frac{123.11 \text{ g } C_6H_5NO_2}{1 \text{ mol } C_6H_5NO_2}$$

$$= 32.00 \text{ g } C_6H_5NO_2$$

Thus, the percentage yield of $C_6H_5NO_2$ is

percentage yield $= \dfrac{28.7 \text{ g}}{32.00 \text{ g}} \times 100\% = 89.69 = 89.7\%$

■ Solutions to Additional Problems

Note on significant figures: The first time the final answer is written, it is written first with one nonsignificant figure. The least significant digit is also underlined. The final answer is then rounded to the correct number of significant figures but the least significant digit is no longer underlined.

8.53 The balanced equation is

$$CaCO_3 + 2\ HCl \longrightarrow CaCl_2 + CO_2 + H_2O$$

The moles of HCl reacted is

$$22.4\ \text{mol}\ CaCO_3 \times \frac{2\ \text{mol}\ HCl}{1\ \text{mol}\ CaCO_3} = 44.8\ \text{mol}\ HCl$$

The moles formed are

$$22.4\ \text{mol}\ CaCO_3 \times \frac{1\ \text{mol}\ CaCl_2}{1\ \text{mol}\ CaCO_3} = 22.4\ \text{mol}\ CaCl_2$$

$$22.4\ \text{mol}\ CaCO_3 \times \frac{1\ \text{mol}\ CO_2}{1\ \text{mol}\ CaCO_3} = 22.4\ \text{mol}\ CO_2$$

$$22.4\ \text{mol}\ CaCO_3 \times \frac{1\ \text{mol}\ H_2O}{1\ \text{mol}\ CaCO_3} = 22.4\ \text{mol}\ H_2O$$

8.54 The balanced equation is

$$Li_2SO_3 + 2\ HBr \longrightarrow 2\ LiBr + SO_2 + H_2O$$

The moles of HBr reacted is

$$2.46\ \text{mol}\ Li_2SO_3 \times \frac{2\ \text{mol}\ HBr}{1\ \text{mol}\ Li_2SO_3} = 4.92\ \text{mol}\ HBr$$

The moles formed are

$$2.46 \text{ mol Li}_2\text{SO}_3 \times \frac{2 \text{ mol LiBr}}{1 \text{ mol Li}_2\text{SO}_3} = 4.92 \text{ mol LiBr}$$

$$2.46 \text{ mol Li}_2\text{SO}_3 \times \frac{1 \text{ mol SO}_2}{1 \text{ mol Li}_2\text{SO}_3} = 2.46 \text{ mol SO}_2$$

$$2.46 \text{ mol Li}_2\text{SO}_3 \times \frac{1 \text{ mol H}_2\text{O}}{1 \text{ mol Li}_2\text{SO}_3} = 2.46 \text{ mol H}_2\text{O}$$

8.55 The balanced equation is

$$\text{Fe}_2\text{O}_3 + 6 \text{ HCl} \longrightarrow 2 \text{ FeCl}_3 + 3 \text{ H}_2\text{O}$$

The moles of HCl reacted is

$$1.23 \text{ mol Fe}_2\text{O}_3 \times \frac{6 \text{ mol HCl}}{1 \text{ mol Fe}_2\text{O}_3} = 7.38 \text{ mol HCl}$$

The moles formed are

$$1.23 \text{ mol Fe}_2\text{O}_3 \times \frac{2 \text{ mol FeCl}_3}{1 \text{ mol Fe}_2\text{O}_3} = 2.46 \text{ mol FeCl}_3$$

$$1.23 \text{ mol Fe}_2\text{O}_3 \times \frac{3 \text{ mol H}_2\text{O}}{1 \text{ mol Fe}_2\text{O}_3} = 3.69 \text{ mol H}_2\text{O}$$

8.56 The balanced equation is

$$3 \text{ H}_2\text{SO}_4 + \text{Al}_2\text{O}_3 \longrightarrow \text{Al}_2(\text{SO}_4)_3 + 3 \text{ H}_2\text{O}$$

The moles of H_2SO_4 required are

$$3.42 \text{ mol Al}_2\text{O}_3 \times \frac{3 \text{ mol H}_2\text{SO}_4}{1 \text{ mol Al}_2\text{O}_3} = 10.26 \text{ mol H}_2\text{SO}_4$$

The moles formed are

$$3.42 \text{ mol Al}_2\text{O}_3 \times \frac{1 \text{ mol Al}_2(\text{SO}_4)_3}{1 \text{ mol Al}_2\text{O}_3} = 3.42 \text{ mol Al}_2(\text{SO}_4)_3$$

$$3.42 \text{ mol Al}_2\text{O}_3 \times \frac{3 \text{ mol H}_2\text{O}}{1 \text{ mol Al}_2\text{O}_3} = 10.26 \text{ mol H}_2\text{O}$$

8.57 The balanced equation is

$$3\ NaOH + H_3PO_4 \longrightarrow Na_3PO_4 + 3\ H_2O$$

Convert the masses to moles. The molar masses of NaOH and H_3PO_4 are 40.00 g and 97.99 g, respectively.

$$5.00\ g\ NaOH \times \frac{1\ mol\ NaOH}{40.00\ g\ NaOH} = 0.1250\ mol\ NaOH$$

$$7.00\ g\ H_3PO_4 \times \frac{1\ mol\ H_3PO_4}{97.99\ g\ H_3PO_4} = 0.07144\ mol\ H_3PO_4$$

Determine the limiting reactant, if there is one. Calculate the amount of H_3PO_4 required to react with 0.1250 mol NaOH.

$$0.1250\ mol\ NaOH \times \frac{1\ mol\ H_3PO_4}{3\ mol\ NaOH} = 0.04167\ mol\ H_3PO_4$$

Since the available H_3PO_4 (0.07144 mol) is more than the required amount, NaOH is the limiting reactant. The molar mass of Na_3PO_4 is 163.94 g. Therefore, the mass of Na_3PO_4 formed is

$$0.1250\ mol\ NaOH \times \frac{1\ mol\ Na_3PO_4}{3\ mol\ NaOH} \times \frac{163.94\ g\ Na_3PO_4}{1\ mol\ Na_3PO_4} = 6.831 = 6.83\ g\ Na_3PO_4$$

The mass of H_3PO_4 that reacts is

$$0.1250\ mol\ NaOH \times \frac{1\ mol\ H_3PO_4}{3\ mol\ NaOH} \times \frac{97.99\ g\ H_3PO_4}{1\ mol\ H_3PO_4} = 4.083\ g\ H_3PO_4$$

Finally, the mass of H_3PO_4 remaining at the end of the reaction is

$$7.00\ g\ H_3PO_4 - 4.083\ g\ H_3PO_4 = 2.917 = 2.92\ g\ H_3PO_4 \qquad \text{(remaining)}$$

8.58 The balanced equation is

$$SO_3 + H_2O \longrightarrow H_2SO_4$$

Convert the masses of reactants to moles. The molar masses of SO_3 and H_2O are 80.06 g and 18.02 g, respectively. Thus,

$$4.56\ g\ SO_3 \times \frac{1\ mol\ SO_3}{80.06\ g\ SO_3} = 0.05696\ mol\ SO_3$$

$$10.00\ g\ H_2O \times \frac{1\ mol\ H_2O}{18.02\ g\ H_2O} = 0.5549\ mol\ H_2O$$

120 ■ CHAPTER 8

Determine the limiting reactant, if there is one. Calculate the amount of H_2O required to react with 0.05696 mol SO_3.

$$0.05696 \text{ mol } SO_3 \times \frac{1 \text{ mol } H_2O}{1 \text{ mol } SO_3} = 0.05696 \text{ mol } H_2O$$

Since the available H_2O (0.5549 mol) is more than the required amount, SO_3 is the limiting reactant. The molar mass of H_2SO_4 is 98.08 g. Thus, the mass of H_2SO_4 formed is

$$0.05696 \text{ mol } SO_3 \times \frac{1 \text{ mol } H_2SO_4}{1 \text{ mol } SO_3} \times \frac{98.06 \text{ g } H_2SO_4}{1 \text{ mol } H_2SO_4} = 5.585 = 5.59 \text{ g } H_2SO_4$$

The mass of H_2O that reacts is

$$0.05696 \text{ mol } SO_3 \times \frac{1 \text{ mol } H_2O}{1 \text{ mol } SO_3} \times \frac{18.02 \text{ g } H_2O}{1 \text{ mol } H_2O} = 1.026 = 1.03 \text{ g } H_2O$$

Thus, the mass of H_2O remaining at the end of the reaction is

$$10.0 \text{ g } H_2O - 1.03 \text{ g } H_2O = 8.97 = 9.0 \text{ g } H_2O \qquad \text{(remaining)}$$

8.59 The balanced equation is

$$Ca(HCO_3)_2 + Ca(OH)_2 \longrightarrow 2 \text{ } CaCO_3 + 2 \text{ } H_2O$$

The molar masses of $CaCO_3$ and $Ca(HCO_3)_2$ are 100.09 g and 162.12 g, respectively. Thus, the mass of $CaCO_3$ formed is

$$16.3 \text{ g } Ca(HCO_3)_2 \times \frac{1 \text{ mol } Ca(HCO_3)_2}{162.12 \text{ g } Ca(HCO_3)_2} \times \frac{2 \text{ mol } CaCO_3}{1 \text{ mol } Ca(HCO_3)_2} \times \frac{100.09 \text{ g } CaCO_3}{1 \text{ mol } CaCO_3}$$

$$= 20.13 = 20.1 \text{ g } CaCO_3$$

8.60 The balanced equation is

$$H_2SO_4 + 2 \text{ NaOH} \longrightarrow Na_2SO_4 + 2 \text{ } H_2O$$

The molar masses of NaOH and Na_2SO_4 are 40.00 g and 142.04 g, respectively. Thus, the mass of Na_2SO_4 formed is

$$2.08 \text{ g NaOH} \times \frac{1 \text{ mol NaOH}}{40.00 \text{ g NaOH}} \times \frac{1 \text{ mol } Na_2SO_4}{2 \text{ mol NaOH}} \times \frac{142.04 \text{ g } Na_2SO_4}{1 \text{ mol } Na_2SO_4}$$

$$= 3.693 = 3.69 \text{ g } Na_2SO_4$$

QUANTITIES IN CHEMICAL REACTIONS ■ 121

8.61 The balanced equation is

$$3\ Ba(OH)_2 + 2\ H_3PO_4 \longrightarrow Ba_3(PO_4)_2 + 6\ H_2O$$

The molar masses of $Ba(OH)_2$ and $Ba_3(PO_4)_2$ are 171.35 g and 601.93 g, respectively. Thus, the mass of $Ba(OH)_2$ required is

$$3.00\ g\ Ba_3(PO_4)_2 \times \frac{1\ mol\ Ba_3(PO_4)_2}{601.93\ g\ Ba_3(PO_4)_2} \times \frac{3\ mol\ Ba(OH)_2}{1\ mol\ Ba_3(PO_4)_2} \times \frac{171.35\ g\ Ba(OH)_2}{1\ mol\ Ba(OH)_2}$$

$$= 2.562 = 2.56\ g\ Ba(OH)_2$$

The molar mass of H_3PO_4 is 97.99 g. Thus, the mass of H_3PO_4 required is

$$3.00\ g\ Ba_3(PO_4)_2 \times \frac{1\ mol\ Ba_3(PO_4)_2}{601.93\ g\ Ba_3(PO_4)_2} \times \frac{2\ mol\ H_3PO_4}{1\ mol\ Ba_3(PO_4)_2} \times \frac{97.99\ g\ H_3PO_4}{1\ mol\ H_3PO_4}$$

$$= 0.9768 = 0.977\ g\ H_3PO_4$$

8.62 The balanced equation is

$$CoCl_2 + 2\ AgNO_3 \longrightarrow Co(NO_3)_2 + 2\ AgCl$$

The molar masses of $CoCl_2$, $AgNO_3$, and $AgCl$ are 129.83 g, 169.88 g, and 143.32 g, respectively. Thus, the masses of $CoCl_2$ and $AgNO_3$ required are

$$1.50\ g\ AgCl \times \frac{1\ mol\ AgCl}{143.32\ g\ AgCl} \times \frac{1\ mol\ CoCl_2}{2\ mol\ AgCl} \times \frac{129.83\ g\ CoCl_2}{1\ mol\ CoCl_2}$$

$$= 0.6794 = 0.679\ g\ CoCl_2$$

$$1.50\ g\ AgCl \times \frac{1\ mol\ AgCl}{143.32\ g\ AgCl} \times \frac{2\ mol\ AgNO_3}{2\ mol\ AgCl} \times \frac{169.88\ g\ AgNO_3}{1\ mol\ AgNO_3}$$

$$= 1.778 = 1.78\ g\ AgNO_3$$

122 ■ CHAPTER 8

8.63 The balanced equations are

$$2\,C + O_2 \longrightarrow 2\,CO$$

$$ZnO + CO \longrightarrow Zn + CO_2$$

First, calculate the moles of CO that can be formed from 50.0 g of carbon. The molar mass of C is 12.01 g. Thus,

$$50.0\text{ g C} \times \frac{1\text{ mol C}}{12.01\text{ g C}} \times \frac{2\text{ mol CO}}{2\text{ mol C}} = 4.163\text{ mol CO}$$

Next, convert the 75.0 g of zinc oxide (ZnO) to moles. The molar mass of ZnO is 81.38 g. Thus,

$$75.0\text{ g ZnO} \times \frac{1\text{ mol ZnO}}{81.38\text{ g ZnO}} = 0.9216\text{ mol ZnO}$$

Next, determine the limiting reactant, if there is one. The amount of CO required to react with 0.9216 mol ZnO is

$$0.9216\text{ mol ZnO} \times \frac{1\text{ mol CO}}{1\text{ mol ZnO}} = 0.9216\text{ mol CO}$$

Since the available CO (4.163 mol) is more than the required amount, ZnO is the limiting reactant. Since the molar mass of zinc is 65.83 g, the mass of Zn formed is

$$0.9216\text{ mol ZnO} \times \frac{1\text{ mol Zn}}{1\text{ mol ZnO}} \times \frac{65.38\text{ g Zn}}{1\text{ mol Zn}} = 60.25 = 60.3\text{ g Zn}$$

8.64 The balanced equations are

$$4\,NH_3 + 5\,O_2 \longrightarrow 4\,NO + 6\,H_2O$$

$$2\,NO + 2\,CH_4 \longrightarrow 2\,HCN + 2\,H_2O + H_2$$

First, calculate the moles of NO that can be formed from 24.2 g of ammonia (NH_3). The molar mass of NH_3 is 17.03 g. Thus,

$$24.2\text{ g NH}_3 \times \frac{1\text{ mol NH}_3}{17.03\text{ g NH}_3} \times \frac{4\text{ mol NO}}{4\text{ mol NH}_3} = 1.421\text{ mol NO}$$

Next, convert the 25.1 g of methane (CH_4) to moles. The molar mass of CH_4 is 16.04 g. Thus,

$$25.1\text{ g CH}_4 \times \frac{1\text{ mol CH}_4}{16.04\text{ g CH}_4} = 1.565\text{ mol CH}_4$$

Next, determine the limiting reactant, if there is one. Calculate the amount of NO required to react with 1.565 mol CH_4.

$$1.565 \text{ mol } CH_4 \times \frac{2 \text{ mol NO}}{2 \text{ mol } CH_4} = 1.565 \text{ mol NO}$$

Since the available NO (1.421 mol) is less than the required amount, NO is the limiting reactant. The molar mass of HCN is 27.03 g. Thus, the mass of HCN formed is

$$1.421 \text{ mol NO} \times \frac{2 \text{ mol HCN}}{2 \text{ mol NO}} \times \frac{27.03 \text{ g HCN}}{1 \text{ mol HCN}} = 38.41 = 38.4 \text{ g HCN}$$

8.65 The balanced equation is

$$C_7H_6O_3 + C_4H_6O_3 \longrightarrow C_9H_8O_4 + HC_2H_3O_2$$

First, convert the masses of reactants to moles. The molar masses of $C_7H_6O_3$ and $C_4H_6O_3$ are 138.12 g and 102.09 g, respectively. Thus,

$$2.00 \text{ g } C_7H_6O_3 \times \frac{1 \text{ mol } C_7H_6O_3}{138.12 \text{ g } C_7H_6O_3} = 0.01448 \text{ mol } C_7H_6O_3$$

$$4.00 \text{ g } C_4H_6O_3 \times \frac{1 \text{ mol } C_4H_6O_3}{102.09 \text{ g } C_4H_6O_3} = 0.03918 \text{ mol } C_4H_6O_3$$

Next, determine the limiting reactant, if there is one. The amount of $C_4H_6O_3$ required to react with 0.01448 mol $C_7H_6O_3$ is

$$0.01448 \text{ mol } C_7H_6O_3 \times \frac{1 \text{ mol } C_4H_6O_3}{1 \text{ mol } C_7H_6O_3} = 0.01448 \text{ mol } C_4H_6O_3$$

Since the available $C_4H_6O_3$ (0.03918 mol) is more than the required amount, $C_7H_6O_3$ is the limiting reactant. The molar mass of aspirin ($C_9H_8O_4$) is 180.15 g. Thus, the mass of aspirin formed is

$$0.01448 \text{ mol } C_7H_6O_3 \times \frac{1 \text{ mol } C_9H_8O_4}{1 \text{ mol } C_7H_6O_3} \times \frac{180.15 \text{ g } C_9H_8O_4}{1 \text{ mol } C_9H_8O_4} = 2.609 = 2.61 \text{ g } C_9H_8O_4$$

8.66 The balanced equation is

$$C_7H_6O_3 + CH_3OH \longrightarrow C_8H_8O_3 + H_2O$$

First, convert the masses of reactants to moles. The molar masses of $C_7H_6O_3$ and CH_3OH are 138.12 g and 32.04 g, respectively. Thus,

$$1.50 \text{ g } C_7H_6O_3 \times \frac{1 \text{ mol } C_7H_6O_3}{138.12 \text{ g } C_7H_6O_3} = 0.01086 \text{ mol } C_7H_6O_3$$

$$11.20 \text{ g } CH_3OH \times \frac{1 \text{ mol } CH_3OH}{32.04 \text{ g } CH_3OH} = 0.3496 \text{ mol } CH_3OH$$

Next, determine the limiting reactant, if there is one. Calculate the amount of CH_3OH required to react with 0.01086 mol $C_7H_6O_3$.

$$0.01086 \text{ mol } C_7H_6O_3 \times \frac{1 \text{ mol } CH_3OH}{1 \text{ mol } C_7H_6O_3} = 0.01086 \text{ mol } CH_3OH$$

Since the available CH_3OH (0.3496 mol) is more than the required amount, $C_7H_6O_3$ is the limiting reactant. The molar mass of $C_8H_8O_3$ is 152.14 g. Thus, the mass of $C_8H_8O_3$ formed is

$$0.01086 \text{ mol } C_7H_6O_3 \times \frac{1 \text{ mol } C_8H_8O_3}{1 \text{ mol } C_7H_6O_3} \times \frac{152.14 \text{ g } C_8H_8O_3}{1 \text{ mol } C_8H_8O_3} = 1.652 = 1.65 \text{ g } C_8H_8O_3$$

9. ELECTRON STRUCTURE OF ATOMS

■ Solutions to Exercises

9.1 There are $3^2 = 9$ orbitals in the n = 3 shell.

9.2 The notation 5d refers to the n = 5 shell and a d-type subshell.

9.3 In the n = 3 shell, there are 3 subshells. Each subshell will begin with 3. They are 3s, 3p, and 3d. The numbers of orbitals in each subshell are 1, 3, and 5, respectively.

9.4 The superscripts show that there are 2 electrons in the 1s subshell, 2 electrons in the 2s subshell, 6 electrons in the 2p subshell, 2 electrons in the 3s subshell, and 5 electrons in the 3p subshell. The total number of electrons is $2 + 2 + 6 + 2 + 5 = 17$.

9.5 Chlorine has atomic number 17 and therefore 17 electrons. The subshells, in order of filling, are 1s, 2s, 2p, 3s, and 3p. Thus, the electron configuration is $1s^2 2s^2 2p^6 3s^2 3p^5$.

9.6 The subshells through germanium (atomic number 32) are 1s, 2s, 2p, 3s, 3p, 4s, 3d, and 4p. Putting the electrons into the orbitals, you get $1s^2 2s^2 2p^6 3s^2 3p^6 4s^2 3d^{10} 4p^2$. It is convenient to rearrange the subshells so that they are in order by shells. This gives the electron configuration $1s^2 2s^2 2p^6 3s^2 3p^6 3d^{10} 4s^2 4p^2$.

9.7 The row number of germanium is 4, so n = 4, and the general form of the valence-shell configuration is $4s^a 4p^b$. The group number is IVA (or 4), so a + b = 4. The valence-shell configuration is $4s^2 4p^2$.

Answers to Questions to Test Your Reading

9.1 When ordinary white light is separated into its components by means of a prism, you obtain a <u>continuous spectrum</u> consisting of a continuous rainbow of colors. But, when you separate the light emitted by heated atoms of an element, you see a <u>line spectrum</u>, or a spectrum consisting of a series of colored lines against a black background. These line spectra are characteristic of a particular kind of atom.

9.2 In Bohr's theory of the atom, it is assumed that the electrons in atoms move in orbits about their nucleus with only certain allowed energies, or <u>energy levels</u>. Also, electrons undergo <u>transitions</u> between energy levels. When electrons undergo a transition to a lower energy level, energy is released as a photon with a particular wavelength, or color.

9.3 An <u>energy level</u> of an atom is one of the allowed energy values that an electron can have, and is designated by a quantum number, n, which is a positive integer.

9.4 The modern theory of atoms is called <u>quantum mechanics</u>.

9.5 An <u>orbit</u> is a specific circular path about the nucleus, much like the earth moves about the sun. An <u>orbital</u> is a region of space around the nucleus where the electron is very likely to be found. It is much less well defined than an orbit. The exact position of the electron in an orbital is not predictable at any given time.

9.6 An <u>electron shell</u>, designated by a shell quantum number, n, is a set of orbitals of approximately the same size and energy. A <u>subshell</u> is a subset of orbitals of an electron shell, each orbital having identical energy and specific shapes. The subshells are denoted s, p, d, f, g, h, and so forth.

9.7 The different possible values of n are any positive whole number value from 1 to infinity.

9.8 For a given shell whose quantum number is n, there are n subshells. They are denoted s, p, d, f, g, h, and so forth, and are preceded by the shell quantum number. For example, a p-type subshell in the n = 2 shell is denoted 2p.

9.9 The number of orbitals in the nth shell of an atom equals n^2. For the n = 5 shell, there should be 5^2 = 25 orbitals. These can be summarized as follows:

subshell	number of orbitals
5s	1
5p	3
5d	5
5f	7
5g	9
Total =	25

The total number of orbitals agrees with the predicted number, 25.

9.10 All s orbitals have a spherical shape, and p orbitals have a dumbbell shape.

9.11 The order (by energy) of the first four subshells of a given shell are s < p < d < f.

9.12 An <u>electron configuration</u> is a particular distribution of electrons among the different subshells of an atom.

9.13 According to the <u>Pauli exclusion principle</u>, an orbital can hold no more than two electrons. Since each subshell has a specified number of orbitals, the maximum number of electrons that a subshell can accommodate is twice the number of orbitals in the subshell. For the n = 1 shell, there is only one subshell (1s) and only one orbital (1s). Thus, the n = 1 shell can accommodate 2 x 1 = 2 electrons. For the n = 2 shell, there are two subshells. The 2s subshell has one orbital, and the 2p subshell has three orbitals. Thus, the n = 2 shell has 1 + 3 = 4 orbitals. Since each orbital can accommodate 2 electrons, the maximum number of electrons that the n = 2 shell can accommodate is 2 x 4 = 8 electrons.

9.14 The first shell (n = 1) can accommodate 2 electrons. Thus, there are two different electron configurations possible for the n = 1 shell. These are $1s^1$ and $1s^2$. Thus, the first row of the periodic table has two elements. The second shell can accommodate 8 electrons. Thus, the second row of the periodic table has eight elements.

9.15 The first three noble gases have the following electron configurations:

noble gas	electron configuration
helium (He)	$1s^2$
neon (Ne)	$1s^2 2s^2 2p^6$
argon (Ar)	$1s^2 2s^2 2p^6 3s^2 3p^6$

9.16 According to the periodic table, the first nine electron subshells in order of filling are: 1s, 2s, 2p, 3s, 3p, 4s, 3d, 4p, 5s.

9.17 <u>Valence electrons</u> are the outer electrons in an atom. They are important in determining the chemical properties of an element.

9.18 Carbon is in Group IVA (or 4). Thus, a + b = 4. The valence-shell electron configuration for a carbon atom is $2s^22p^2$.

Solutions to Practice Problems

9.19 Since shorter wavelengths correspond to higher frequencies, violet light, with a wavelength of 400 nm, will have the higher frequency.

9.20 Since shorter wavelengths correspond to higher frequencies, red light, with a wavelength of 700 nm, will have the higher frequency.

9.21 Since shorter wavelengths correspond to higher energies, green light, with a wavelength of 500 nm, will have the higher energy..

9.22 Since shorter wavelengths correspond to higher energies, X rays, with a wavelength of 1 nm, will have the higher energy.

9.23 Since higher values of n correspond to higher energy, the n = 4 energy level has the greater energy.

9.24 Since higher values of n correspond to higher energy, the n = 6 energy level has the greater energy.

9.25 For both hydrogen atoms, the electron starts in the same energy level (n = 5). The electron that undergoes a transition to a lower final energy state will release more energy than a transition to a higher final energy state. Thus, the transition from the n = 5 level to the n = 2 level will emit a photon with greater energy.

9.26 For both hydrogen atoms, the electron starts in the same energy level (n = 5). The electron that undergoes the transition to a higher final energy state will release less energy than a transition to a lower final energy state. Thus, the transition from the n = 5 level to the n = 4 level will emit a photon with lesser energy.

9.27 a. $2^2 = 4$ b. $5^2 = 25$

9.28 a. $1^2 = 1$ b. $6^2 = 36$

9.29 a. The notation 3p refers to the n = 3 shell and a p-type subshell.

b. The notation 4d refers to the n = 4 shell and a d-type subshell.

c. The notation 2s refers to the n = 2 shell and a s-type subshell.

ELECTRON STRUCTURE OF ATOMS ■ 129

9.30 a. The notation 4f refers to the n = 4 shell and an f-type subshell.

b. The notation 3s refers to the n = 3 shell and an s-type subshell.

c. The notation 5p refers to the n = 5 shell and a p-type subshell.

9.31 Because n = 2, there are 2 subshells, 2s and 2p. The number of orbitals in each subshell are 1 and 3, respectively.

9.32 Because n = 5, there are 5 subshells, 5s, 5p, 5d, 5f, and 5g. The number of orbitals in each subshell are 1, 3, 5, 7, and 9, respectively.

9.33 3s < 3p < 3d.

9.34 4s < 4p < 4d < 4f.

9.35 The superscripts show that there are 2 electrons in the 1s subshell, 2 electrons in the 2s subshell, 6 electrons in the 2p subshell, 2 electrons in the 3s subshell, and 4 electrons in the 3p subshell. The total number of electrons in the atom is 2 + 2 + 6 + 2 + 4 = 16.

9.36 The superscripts show that there are 2 electrons in the 1s subshell, 2 electrons in the 2s subshell, 6 electrons in the 2p subshell, 2 electrons in the 3s subshell, 6 electrons in the 3p subshell, 5 electrons in the 3d subshell, and 2 electrons in the 4s subshell. The total number of electrons in the atom is 2 + 2 + 6 + 2 + 6 + 5 + 2 = 25.

9.37 The n = 2 shell can hold 2×2^2 = 8 electrons.

9.38 The n = 3 shell can hold 2×3^2 = 18 electrons.

9.39 The electron configuration for a silicon atom is $1s^2 2s^2 2p^6 3s^2 3p^2$.

9.40 The electron configuration for a magnesium atom is $1s^2 2s^2 2p^6 3s^2$.

9.41 The electron configuration for an arsenic atom is $1s^2 2s^2 2p^6 3s^2 3p^6 3d^{10} 4s^2 4p^3$.

9.42 The electron configuration for a vanadium atom is $1s^2 2s^2 2p^6 3s^2 3p^6 3d^3 4s^2$.

9.43 Strontium (Sr) is in Group IIA (or 2) and period 5. Thus, the valence-shell configuration is $5s^2$.

9.44 The valence-shell configuration for a bismuth atom is $6s^2 6p^3$.

9.45 The valence-shell configuration for this atom is $4s^2 4p^4$. Thus, it is in period 4 and Group VIA. The element is selenium (Se).

9.46 The valence-shell configuration for this atom is $4s^24p^6$. Thus, it is in period 4 and Group VIII. The element is krypton (Kr).

■ Solutions to Additional Problems

9.47 a. 13   b. 5   c. 18   d. 35

9.48 a. 24   b. 29   c. 9   d. 19

9.49 a. 2p is allowed.

b. 2d is not allowed. The n = 2 shell has only 2 subshells, 2s and 2p.

c. 3f is not allowed. The n = 3 shell has only 3 subshells, 3s, 3p, and 3d.

d. 4f is allowed.

9.50 a. 1p is not allowed. The n = 1 shell has only 1 subshell, 1s.

b. 3d is allowed.

c. 4d is allowed.

d. 0s is not allowed. The value of n = 0 is not allowed.

9.51 a. lowest energy

b. lowest energy.

c. not allowed.

d. lowest energy.

9.52 a. not lowest energy.

b. lowest energy.

c. lowest energy.

d. not allowed.

9.53 The first four elements of Group IVA have the following electron configurations:

element	electron configuration
carbon (C)	$1s^22s^22p^2$
silicon (Si)	$1s^22s^22p^63s^23p^2$
germanium (Ge)	$1s^22s^22p^63s^23p^63d^{10}4s^24p^2$
tin (Sn)	$1s^22s^22p^63s^23p^63d^{10}4s^24p^64d^{10}5s^25p^2$

9.54 The first four elements of Group VIA have the following electron configurations:

element	electron configuration
oxygen (O)	$1s^22s^22p^4$
sulfur (S)	$1s^22s^22p^63s^23p^4$
selenium (Se)	$1s^22s^22p^63s^23p^63d^{10}4s^24p^4$
tellurium (Te)	$1s^22s^22p^63s^23p^63d^{10}4s^24p^64d^{10}5s^25p^4$

9.55 The total number of electrons in the atom, and thus the atomic number, is $2 + 2 + 6 + 2 + 2 + = 22$. This corresponds to the element titanium (Ti).

9.56 The total number of electrons in the atom, and thus the atomic number, is $2 + 2 + 6 + 2 + 6 + 10 + 2 + 5 + = 35$. This corresponds the element bromine (Br).

9.57 a. 1   b. 2   c. 4   d. 6

9.58 a. 7   b. 3   c. 1   d. 5

9.59 The first four elements of Group IIIA have the following valence-shell configurations:

element	valence-shell configuration
boron (B)	$2s^22p^1$
aluminum (Al)	$3s^23p^1$
gallium (Ga)	$4s^24p^1$
indium	$5s^25p^1$

9.60 The first four elements of Group IIIA have the following valence-shell configurations:

element	valence-shell configuration
nitrogen (N)	$2s^22p^3$
phosphorus (P)	$3s^23p^3$
arsenic (As)	$4s^24p^3$
antimony (Sb)	$5s^25p^3$

10. PERIODIC PROPERTIES OF THE ELEMENTS

■ Solutions to Exercises

10.1 Since P and S are both in Group VA, As is more metallic.

10.2 Since O and Se are both in Group VIA, O is more nonmetallic.

10.3 Since O and F are in the same period, F is the smaller atom.

10.4 Since F^- and Cl^- are in the same group (Group VIIA), F^- is smaller.

10.5 The atom with the smaller ionization energy is O.

■ Answers to Questions to Test Your Reading

10.1 In general, the elements tend toward more nonmetallic character as you move left to right in the periodic table. For example, in the second period, lithium (Li) and beryllium (Be) are metals, boron (B) is a metalloid, and the remainder from carbon (C) to neon (Ne) are nonmetals.

10.2 In general, the elements tend toward more metallic character as you move from the top of a group to the bottom. For example, Group IVA begins with the nonmetal carbon (C), then two metalloids, silicon (Si) and germanium (Ge), and then two metals, tin (Sn) and lead (Pb).

10.3 Chemically, the property of an element that makes it a metal is the element's ability to lose its valence electrons to form a cation. On the other hand, a nonmetal has the ability to gain electrons and form an anion.

10.4 $2\ K(s)\ +\ 2\ H_2O(l)\ \longrightarrow\ 2\ KOH(aq)\ +\ H_2(g)$

10.5 Since the alkali metals are very reactive with water, they are stored in kerosene to prevent contact with moist air.

10.6 The Group IA elements are known as the <u>alkali metals</u> because the reaction of the metal with water produces an alkaline (basic) solution.

10.7 Barium is more reactive than magnesium. As expected, the reactivity of the metal increases as you go down in a given group.

10.8 $Ba(s) + 2 H_2O(l) \longrightarrow Ba(OH)_2(aq) + H_2(g)$

10.9 The gaseous substance that forms when aluminum metal is added to a beaker of acid is hydrogen gas (H_2).

10.10 The two forms of carbon are graphite, a black solid, and diamond, a hard colorless solid.

10.11 Carbon compounds that are burned in air form carbon dioxide (CO_2) and some carbon monoxide (CO).

10.12 Since carbon is a nonmetal and lead is a metal, we would not expect carbon to form a compound similar to $PbSO_4$.

10.13 Nitrogen gas (N_2) protects the hot filament in a light bulb from reacting with the oxygen in the air.

10.14 White phosphorus is stored under water because it is very reactive with oxygen present in air.

10.15 The two most reactive elements in Group VIA are oxygen and sulfur. An example of a reaction is the burning of sulfur in air to form sulfur dioxide.

$S_8(s) + 8 O_2(g) \longrightarrow 8 SO_2(g)$

10.16 Molecular oxygen is a diatomic molecule, O_2. On the other hand, molecular sulfur has eight atoms, S_8.

10.17 Fluorine (F_2) is a pale yellow gas, chlorine (Cl_2) is a pale greenish-yellow gas, bromine (Br_2) is a reddish-brown liquid, and iodine (I_2) is a bluish-black solid.

10.18 $Mg(s) + Br_2(l) \longrightarrow MgBr_2(s)$

10.19 We should expect greater reactivity for xenon, krypton, and radon since their valence electrons become increasingly farther from the nucleus as you go down in the group.

10.20 First, the <u>atomic radius</u> increases in any given main-group as you move down the group of elements. Second, the atomic radius decreases in any given period of main-group elements as you move across the period.

10.21 Since barium is below strontium, it has an additional shell of electrons and therefore a larger atomic radius.

10.22 For an isoelectronic series of ions, the ionic radius decreases as the atomic number increases.

10.23 K + energy $\longrightarrow$ K$^+$ + e$^-$

10.24 Since strontium and barium are both in Group IIA, the ionization energy will decrease. This is because the atoms are getting bigger, and the electrons are not held as tightly. Thus, they can be removed easier.

10.25 <u>Electron affinity</u> refers to how strongly an atom tends to gain electrons from other atoms.

10.26 The ionization energies of nonmetals are larger than the ionization energies of metals. This is because metals tend to lose electrons easily while nonmetals tend to gain electrons easily.

10.27 The strongest electron affinities are for the elements fluorine, oxygen, and chlorine.

10.28 Noble gases have the largest ionization energies and also very small electron affinities. This makes them unreactive.

■ Solutions to Practice Problems

10.29 a. Sn and Sr are both in the same period and thus, Sr is more metallic. It is farther to the left in the period.

  b. S and Te are both in Group VIA and thus, Te is more metallic. It is closer to the bottom of the group.

10.30 a. Bi and Ba are both in the same period and thus, Ba is more metallic. It is farther to the left in the period.

  b. Li and Cs are both in Group IA and thus, Cs is more metallic. It is closer to the bottom of the group.

PERIODIC PROPERTIES OF THE ELEMENTS ■ 135

10.31 a. P and N are both in Group VA and thus, N is more nonmetallic. It is closer to the top of the group.

b. Se and As are both in the same period and thus, Se is more nonmetallic. It is farther to the right in the period.

10.32 a. C and Ge are both in Group IVA and thus, C is more nonmetallic. It is closer to the top of the group.

b. S and Si are both in the same period and thus, S is more nonmetallic. It is farther to the right in the period.

10.33   a. Br is larger than F.   b. K is larger than Br.

10.34   a. Te is larger than S.   b. Rb is larger than I.

10.35   a. O^{2-} is smaller than Se^{2-}.   b. Cl^- is smaller than S^{2-}.

10.36   a. Ga^{3+} is smaller than S^{2-}.   b. Ca^{2+} is smaller than Ba^{2+}.

10.37 a. F has a greater ionization energy than Br.

b. Br has a greater ionization energy than K.

10.38 a. S has a greater ionization energy than Te.

b. I has a greater ionization energy than Rb.

■ Solutions to Additional Problems

10.39   Be

10.40   F

10.41   Li is the largest atom in period 2.

10.42   Ar is the smallest atom in period 3.

10.43   Ca is more metallic than Be.

10.44   I is more metallic than F.

10.45 The least metallic element in Group IIIA is B.

10.46 The least metallic element in Group VIA is O.

10.47 In order of increasing atomic radius: F < Cl < S.

10.48 In order of increasing atomic radius: S < Se < As.

10.49 In order of decreasing ionization energy: S > Mg > Sr.

10.50 In order of decreasing ionization energy: Cl > Al > Na.

10.51 In order of increasing size: $F^- < Cl^- < Br^- < I^-$.

10.52 In order of increasing size: $Be^{2+} < Mg^{2+} < Ca^{2+} < Ba^{2+}$.

10.53 In order of decreasing size: $Se^{2-} > Br^- > Rb^+ > Sr^{2+}$.

10.54 In order of decreasing size: $P^{3-} > S^{2-} > Cl^- > K^+$.

10.55 Na is larger than Na^+.

10.56 Al is larger than Al^{3+}.

10.57 Cl is smaller than Cl^-.

10.58 Se is smaller than Se^{2-}.

11. CHEMICAL BONDING

■ Solutions to Exercises

11.1 a. The Al atom is the metal and is in group IIIA. The number of electrons lost by Al is 3, equal to the group number. The F atom is the nonmetal and is in Group VIIA. The number of electrons picked up by F is 8 - 7 = 1. Therefore the ions formed are Al^{3+} and F^-. Since the Al atom loses three electrons and the F atoms pick up one electron, you need one Al atom and three F atoms. Using electron-dot symbols, the equation for the formation of ions is

$$\cdot Al \cdot \;+\; 3 \;\cdot \ddot{\underset{..}{F}}: \;\longrightarrow\; Al^{3+} \;+\; 3 \left[:\ddot{\underset{..}{F}}:\right]^-$$

Each of the resulting ions has a noble-gas electron configuration.

b. Calcium is a metal and is in Group IIA. Thus it loses two electrons and forms the Ca^{2+} cation. Oxygen is a nonmetal and is in Group VIA. Thus it picks up 8 - 6 = 2 electrons and forms the O^{2-} anion. In simplest whole numbers the ratio of Ca^{2+} to O^{2-} is 1 to 1. The resulting equation for the formation of ions is

$$\cdot Ca \cdot \;+\; \cdot \ddot{\underset{..}{O}}: \;\longrightarrow\; Ca^{2+} \;+\; \left[:\ddot{\underset{..}{O}}:\right]^{2-}$$

The resulting ions have a noble-gas electron configuration.

11.2

a. The oxygen atom in water, H_2O, is the central atom. The arrangement of the atoms in the molecule is

    H   O   H

<u>Step 1.</u> The total number of valence electrons is obtained as follows: O is in Group VIA, so has 6 valence electrons; each H has 1 valence electron. So the total number of valence electrons is $1(6) + 2(1) = 8$.
<u>Step 2.</u> Connect the atoms with a pair of dots.

    H : O : H

<u>Step 3.</u> Counting up the electrons used so far you get 4, which leaves $8 - 4 = 4$ electrons remaining (2 electron pairs).
<u>Step 4.</u> Place the electron pairs on the central atom.

              ..
    H : O : H
              ..

Each atom satisfies the octet rule. Therefore the final electron-dot formula is

              ..   ..
    H : O : H   or   H ─── O ─── H
              ..   ..

b. $HClO_3$, chloric acid, is an oxyacid. You place the Cl atom in the center with the O atoms around it and you bond the H atom to one of the O atoms.

            O

    O   Cl   O   H

<u>Step 1.</u> Hydrogen has 1 valence electron, oxygen has 6 valence electrons and chlorine has 7. The total number of valence electrons is $1(1) + 3(6) + 1(7) = 26$
<u>Step 2.</u> Connect the atoms by electron pairs

              O
              ..
    O : Cl : O : H

<u>Step 3.</u> Distribute electron pairs to the O atoms to satisfy the octet rule.

              ..
            : O :
        ..   ..   ..
     : O : Cl : O : H
        ..   ..

24 electrons have been used so far so there are $26 - 24 = 2$ electrons remaining (1 electron pair).

CHEMICAL BONDING ■ 139

Step 4. Place the electron pair on the central atom.

$$:\!\ddot{\underset{..}{O}}\!:$$
$$:\!\ddot{\underset{..}{O}}\!:\!\ddot{\underset{..}{Cl}}\!:\!\ddot{\underset{..}{O}}\!:\!H$$

Each atom satisfies the octet rule. Therefore the final electron-dot formula is

$$:\!\ddot{\underset{..}{O}}\!:\!\ddot{\underset{..}{Cl}}\!:\!\ddot{\underset{..}{O}}\!:\!H \quad \text{or} \quad :\!\ddot{\underset{..}{O}}\!-\!\ddot{\underset{..}{Cl}}\!-\!\ddot{\underset{..}{O}}\!-\!H$$
(with $:\ddot{O}:$ above Cl in both)

11.3 a. Nitrogen, N_2, is a diatomic molecule. The arrangement of atoms is

N   N

Step 1. Each N atom has 5 valence electrons (Group VA) for a total of 2(5) = 10.
Step 2. Connect the atoms with an electron pair

N : N

Step 3. Add electron pairs to one N atom to make an octet. Place the remaining electron pair on the other N atom

$$:\!\ddot{\underset{..}{N}}\!:\!N\!:$$

This uses all 10 electrons available.
Step 4. Note that the second N needs two electron pairs to satisfy its octet. Move these electron pairs from the other N into the bonding region to form a triple bond

: N ::: N :   or   : N≡N :

Note that all atoms have octets.

b. The carbon atom in carbon disulfide, CS_2, is more electropositive than the S atoms so it is the central atom. The arrangement of atoms is

S   C   S

Step 1. C has 4 valence electrons (Group IVA) and sulfur has 6 valence electrons (Group VIA). The total number of valence electrons is 1(4) + 2(6) = 16.

Step 2. Connect the atoms by electron pairs

$$S : C : S$$

Step 3. Add electron pairs to the S atoms to give them an octet

$$:\overset{..}{\underset{..}{S}} : C : \overset{..}{\underset{..}{S}} :$$

This uses all 16 electrons available.
Step 4. Note that the C atom requires two more electron pairs to have an octet. Move one pair of electrons from each S atom to the bonding regions, giving two double bonds

$$\overset{..}{\underset{..}{S}} :: C :: \overset{..}{\underset{..}{S}} \quad \text{or} \quad \overset{..}{\underset{..}{S}} = C = \overset{..}{\underset{..}{S}}$$

Note that all atoms have octets.

11.4 The boron atom in the tetrafluoroborate ion, BF_4^-, is more electropositive than the F atoms so it is the central atom. The arrangement of atoms is

```
        F
    F   B   F
        F
```

Step 1. B has 3 valence electrons (Group IIIA) and F has 7 valence electrons (Group VIIA). You need to add one electron to account for the negative charge on BF_4^-. Therefore, the total number of valence electrons is $1(3) + 4(7) + 1 = 32$.
Step 2. Connect the atoms by electron pairs

```
        F
        ..
    F : B : F
        ..
        F
```

Step 3. Distribute the electron pairs around the F atoms to give them octets This uses 32 electrons, which is all that is available.

$$\begin{array}{c} ..\\ :F:\\ ..\\ ..\quad ..\quad ..\\ :F:B:F:\\ ..\quad\quad..\\ :F:\\ .. \end{array}$$

Step 4. Note that all atoms have octets. Therefore, the electron-dot formula for the ion is

$$\left[\begin{array}{c} ..\\ :F:\\ ..\quad ..\quad ..\\ :F:B:F:\\ ..\quad\quad..\\ :F:\\ .. \end{array}\right]^- \quad \text{or} \quad \left[\begin{array}{c} ..\\ :F:\\ ..\ |\ ..\\ :F-B-F:\\ ..\ |\ ..\\ :F:\\ .. \end{array}\right]^-$$

11.5 a. The electron-dot formula of the ammonium ion, NH_4^+, is

$$\left[\begin{array}{c} H\\ ..\\ H:N:H\\ ..\\ H \end{array}\right]^+$$

Note that there are four electron pairs about the N atom. These four pairs have a tetrahedral arrangement. Thus the ion has a tetrahedral shape.

b. The electron-dot formula of nitrogen trifluoride, NF_3, is

$$\begin{array}{c} ..\quad ..\quad ..\\ :F:N:F:\\ ..\quad ..\quad ..\\ :F:\\ .. \end{array}$$

There are four electron pairs about the N atom, and these four pairs have a tetrahedral arrangement. Looking at just the atoms and bonds, you would predict that the molecule has a triangular pyramidal shape.

Answers to Questions to Test Your Reading

11.1 The two principal types of chemical bonds are <u>covalent</u> and <u>ionic</u>.

11.2 An <u>ionic bond</u> is the strong attractive force that exists between a positive ion and a negative ion in an ionic compound.

11.3 The forces that are involved in ionic bonding are the electrical forces of attraction between the positive and negative ions.

11.4 The lithium atom has one valence electron, which is easily lost. The ionization process can be described as follows:

$$\text{Li atom } (1s^22s^1) + \text{ionization energy} \longrightarrow \text{Li}^+ \text{ ion } (1s^2) + e^-$$

The resulting lithium ion has the same electron configuration as the noble gas helium.

The fluorine atom has seven valence electrons. It tends to gain an electron to form the fluoride ion, F^-, releasing energy in the process. The ionization process can be described as follows:

$$\text{F atom } (1s^22s^22p^5) + e^- \longrightarrow \text{F}^- \text{ ion } (1s^22s^22p^6) + \text{energy}.$$

The added electron goes into the 2p subshell, giving the fluoride ion an electron configuration the same as that of the noble gas neon.

11.5 The energy term that is associated with the formation of cations from atoms is the ionization energy. The corresponding term for the formation of anions from atoms is the electron affinity.

11.6 Ionic bonds are formed between atoms of metal elements with atoms of nonmetal elements.

11.7 For a main-group element, a metal forms a cation, with a charge equal to the group number. Nonmetal atoms form anions with charge equal to the group number minus 8.

11.8 a. K and O should form an ionic bond. (metal + nonmetal)

b. C and O should form a covalent bond. (two nonmetals)

c. Ca and Cl should form an ionic bond. (metal + nonmetal)

d. N and F should form a covalent bond. (two nonmetals)

CHEMICAL BONDING ■ 143

11.9 a. Cs will form the Cs⁺ ion. (atoms in Group IA have one valence electron)

b. Ba will form the Ba^{2+} ion. (atoms in Group IIA have two valence electrons)

c. S will form the S^{2-} ion. (atoms in Group VIA will gain 8 - 6 = 2 electrons)

d. I will form the I^- ion. (atoms in Group VIIA will gain 8 - 7 = 1 electron)

11.10 The <u>octet rule</u> says that atoms tend to lose or gain electrons when bonding to give eight electrons in their valence shells (except that elements of lowest atomic number tend to lose or gain electrons to give a single shell of two electrons as in the noble gas helium).

11.11 (a) Rb ·   (b) · Ba ·   (c) : Se ·   (d) : I ·

11.12 A <u>covalent bond</u> is formed by the sharing of a pair of electrons between two atoms. This sharing of electrons results in an octet of electrons in the valence shells of the atoms. It is different from an <u>ionic bond</u> in that the atoms do not lose or gain electrons in achieving their octets. Instead, a molecular orbital forms in which the electrons occupy the region around both atoms.

11.13 An <u>electron dot symbol</u> is a symbol of an atom or ion. An <u>electron dot formula</u> represents covalent bonds between atoms in a molecule.

11.14 : Cl · + · Cl : ⟶ : Cl : Cl : or : Cl — Cl :

11.15 a. <u>single bond:</u>   A · + · B ⟶ A : B

b. <u>double bond:</u>   : O · + · O : ⟶ O : : O   or   O = O

c. Lone pair
H — N — H
  |
  H

11.16 A <u>multiple bond</u> is a double or triple bond that is formed by sharing two pairs or three pairs of electrons, respectively, between atoms in a covalent bond.

11.17 Electronegativity is a measure of an atom in a covalent bond to draw bonding electrons to itself.

11.18 The basic trends in electronegativity are first, it increases in any row (period) from left to right and second, it decreases in any column (group) from top to bottom.

11.19 a. As (same period)   b. S (same group)
         c. Br (same period)   d. F (same group)

11.20 (a) $\overset{\delta^+}{C}$—$\overset{\delta^-}{O}$ (b) $\overset{\delta^+}{H}$—$\overset{\delta^-}{S}$ (c) $\overset{\delta^-}{Cl}$—$\overset{\delta^+}{C}$ (d) $\overset{\delta^-}{F}$—$\overset{\delta^+}{N}$

11.21 a. In H_2Te, Te is the central atom since hydrogen cannot be a central atom.

b. In CCl_4, the carbon atom is more electropositive than chlorine so carbon is the central atom.

c. In OF_2, the oxygen atom is more electropositive than fluorine so oxygen is the central atom.

d. HClO is an oxyacid. The oxygen atom is bonded to the chlorine and the hydrogen atom is bonded to the oxygen. This makes oxygen the central atom. Note that in all oxyacids with more than one oxygen atom, the atom that is not hydrogen or oxygen is the central atom.

11.22 a. In BF_4^-, boron (Group IIIA) has 3 valence electrons and fluorine (Group VIIA) has 7. You must add one electron for the negative charge on the ion. The total number of valence electrons is 1(3) + 4(7) + 1 = 32.

b. In SCl_2, sulfur (Group VIA) has 6 valence electrons and chlorine (Group VIIA) has 7. The total number of valence electrons is 1(6) + 2(7) = 20.

c. In H_2SO_4, hydrogen (Group IA) has 1 valence electron, sulfur (Group VIA) has 6, and oxygen (Group VIA) has 6. The total number of valence electorns is 2(1) + 1(6) + 4(6) = 32.

d. In PCl_3, phosphorus (Group VA) has 5 valence electrons and chlorine (Group VII) has 7. The total number of valence electrons is 1(5) + 3(7) = 26.

11.23 Bond length is the normal distance between nucleii whose atoms form a bond in a molecule.

11.24 To describe molecular structure you must specify the bond lengths and bond angles.

CHEMICAL BONDING ■ 145

11.25 VSEPR, or valence shell electron pair repulsion, is a model for predicting the shapes of molecules and ions, in which the valence-shell electron pairs are arranged about each atom in such a way as to keep electron pairs as far away from one another as possible.

11.26 A molecule will be polar if, first, the component elements have a difference in electronegativities and, second, the central atom has lone pairs of electrons.

■ **Solutions to Practice Problems**

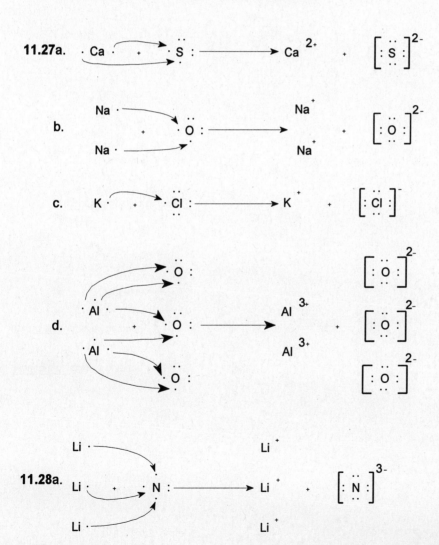

b.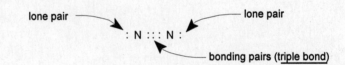

c. (reaction showing Li· + ·Br: → Li⁺ + [:Br:]⁻)

d. (reaction showing 2 Li· + ·S: → 2 Li⁺ + [:S:]²⁻)

11.29 a. :N≡N:

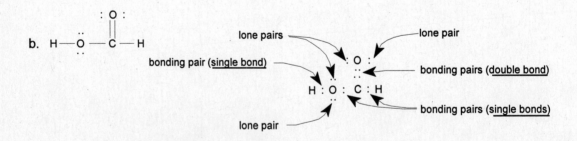

b. H—Ö—C(=Ö:)—H

CHEMICAL BONDING ■ 147

11.30 a.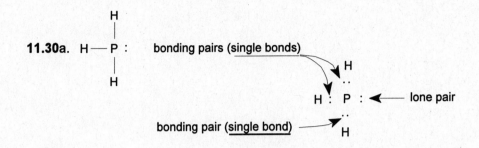

b.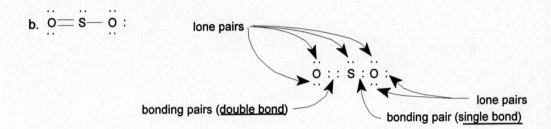

11.31 a. 4.0 - 0.9 = 3.1   b. 3.0 - 2.5 = 0.5   c. 4.0 - 2.5 = 1.5   d. 2.5 - 2.5 = 0
Order of bonds (pure covalent to ionic): C–C, C–Cl, F–C, F–Na

11.32 a. 4.0 - 1.0 = 3.0   b. 3.0 - 3.0 = 0   c. 4.0 - 3.0 = 1.0   d. 3.0 - 3.0 = 0
Order of bonds (pure covalent to ionic): N–N, Cl–N, F–N, Li–F

11.33 a. :F̈:
           :F̈—Ge—F̈:
              :F̈:

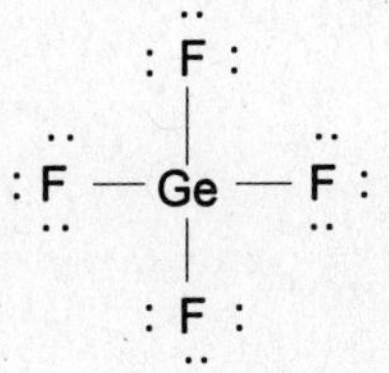

b. :F̈—As—F̈:
        :F̈:

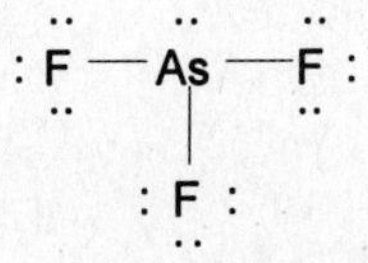

c. H—As̈—H
         |
         H

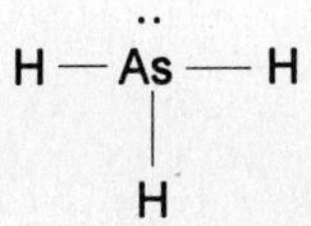

d. :C̈l—Ö—H

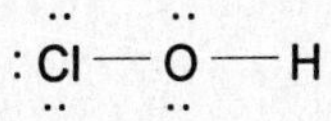

11.34 a. 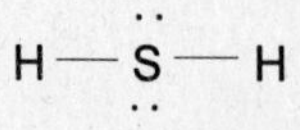 H—S̈—H b. 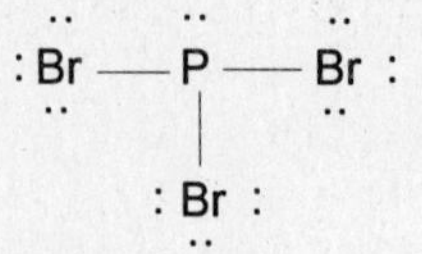 :Br̈—P—Br̈:
                      |
                     :Br̈:

c. 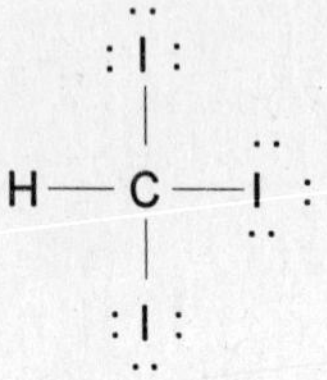 H—C(—Ï:)(—Ï:)—Ï:
d. 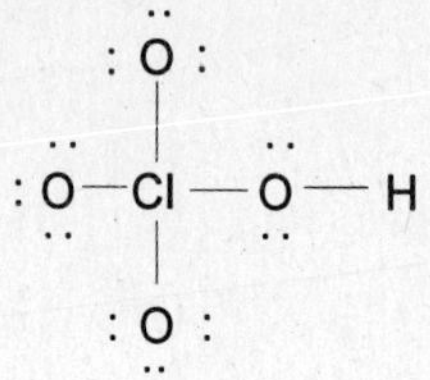 :Ö—Cl(=Ö)(—Ö:)—Ö—H

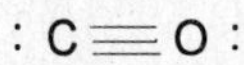

11.35 a. 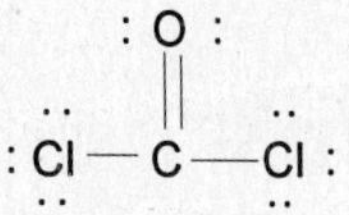 :Cl̈—C(=Ö)—Cl̈:   b. :C≡O: c. :Br̈—C≡N:

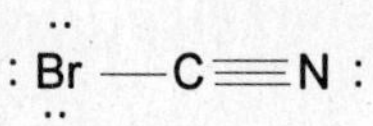

11.36 a. 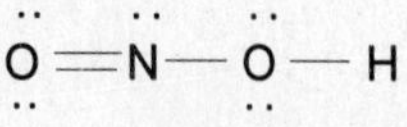 Ö=N̈—Ö—H   b. 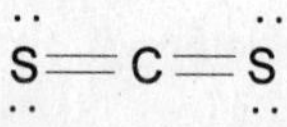 S̈=C=S̈   c. S̈=N̈—F̈:

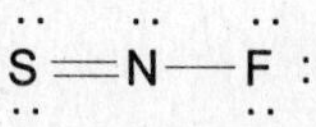

11.37 a. 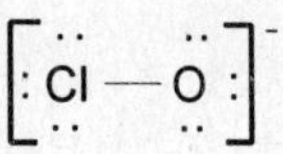 [:Cl̈—Ö:]⁻   b. 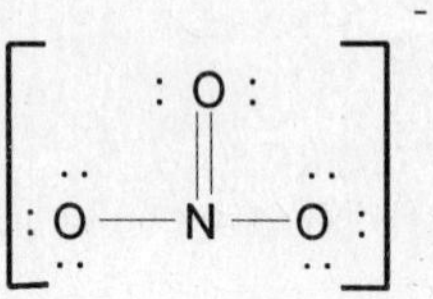 [:Ö—N(=Ö)—Ö:]⁻

11.38 a. 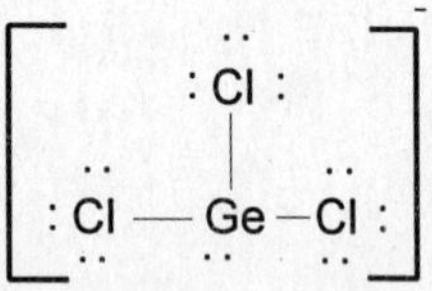 [:Cl̈—Ge(—Cl̈:)(—Cl̈:)]⁻   b. 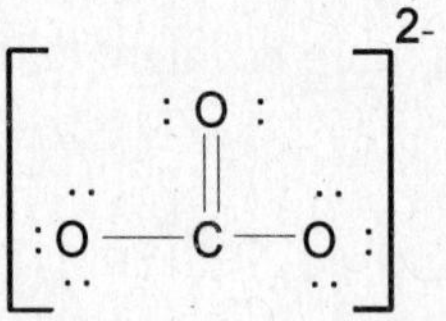 [:Ö—C(=Ö)—Ö:]²⁻

11.39 a. Tetrahedral shape

b. Triangular pyramidal shape

c. Trigonal planar shape

11.40 a. Bent or angular shape

b. Bent or angular shape

c. Tetrahedral shape

Solutions to Additional Problems

11.41 LiCl is ionic (metal + nonmetal) and Cl_2 is covalent (two nonmetals). The formation of the ionic bond in LiCl can be described as follows:

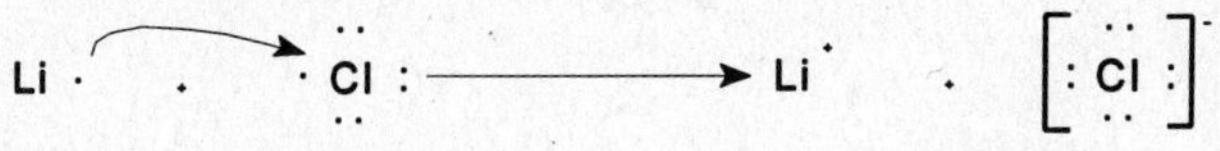

Li· + ·C̈l: ⟶ Li⁺ + [:C̈l:]⁻

The formation of the covalent bond in Cl_2 is as follows:

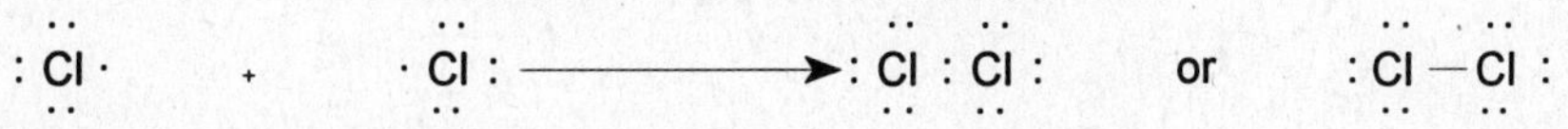

:C̈l· + ·C̈l: ⟶ :C̈l:C̈l: or   :C̈l—C̈l:

11.42 CsBr is ionic (metal + nonmetal) and Br_2 is covalent (two nonmetals). The formation of the ionic bond in CsBr can be described as follows:

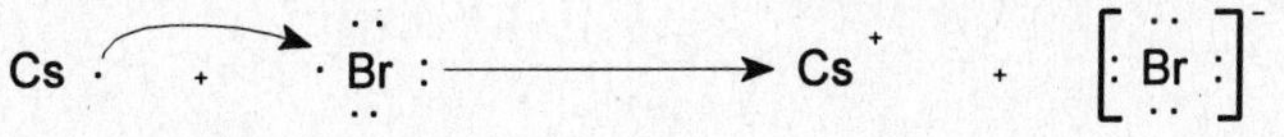

Cs· + ·B̈r: ⟶ Cs⁺ + [:B̈r:]⁻

The formation of the covalent bond in Br_2 is as follows:

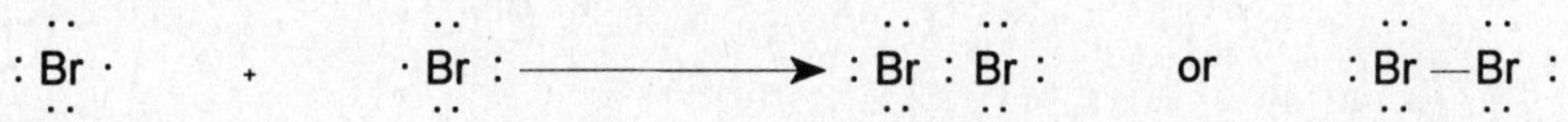

:B̈r· + ·B̈r: ⟶ :B̈r:B̈r: or   :B̈r—B̈r:

11.43 The electron configuration of the Pb^{2+} ion is $[Xe]4f^{14}5d^{10}6s^2$.

11.44 The electron configuration of the Sn^{2+} ion is $[Kr]4d^{10}5s^2$.

11.45 a. Na–O   b. H–O   c. F–H

11.46 a. H–F   b. Li–Cl   c. N–Cl

11.47 a.   Na$^+$ + $[:\!\ddot{\underset{..}{Br}}\!:]^-$

b.   Ca^{2+} + $[:\!\ddot{\underset{..}{F}}\!:]^-$
                         $[:\!\ddot{\underset{..}{F}}\!:]^-$

c.   Na$^+$ Na$^+$ + $[:\!\ddot{\underset{..}{S}}\!:]^{2-}$

11.48 a.   Mg^{2+} + $[:\!\ddot{\underset{..}{Br}}\!:]^-$
                         $[:\!\ddot{\underset{..}{Br}}\!:]^-$

b.   Ca^{2+} + $[:\!\ddot{\underset{..}{O}}\!:]^{2-}$

c.   Na$^+$ Na$^+$ + $[:\!\ddot{\underset{..}{O}}\!:]^{2-}$

11.49 a. 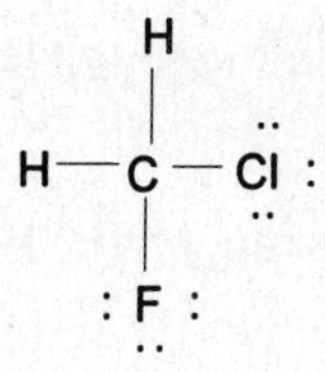 H—C(H)(Cl)—F   (H on top, :Cl: right, :F: bottom, C center)

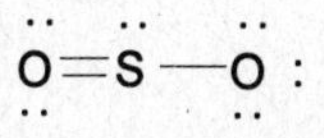

b. :Ö=S—Ö:

c. :Ö—N(=O:)—O—H

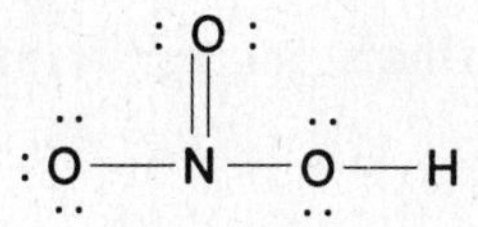

d. 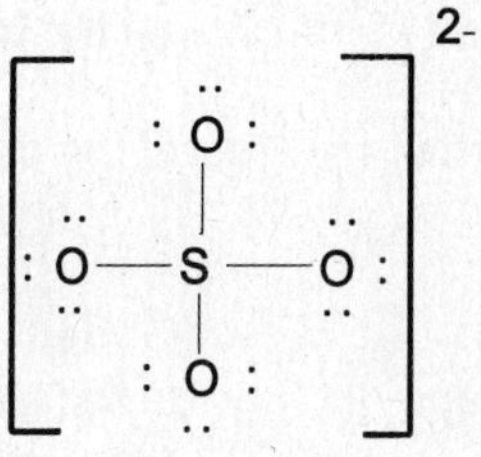 [:Ö—S(=O:)(=O:)—Ö:]$^{2-}$

11.50 a. 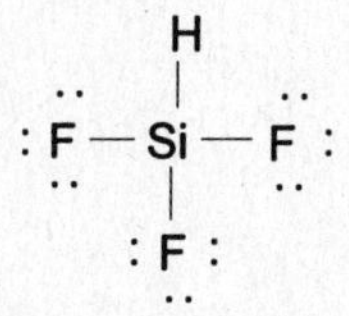 :F̈—Si(F:)—F̈:   (three F around Si)

b. 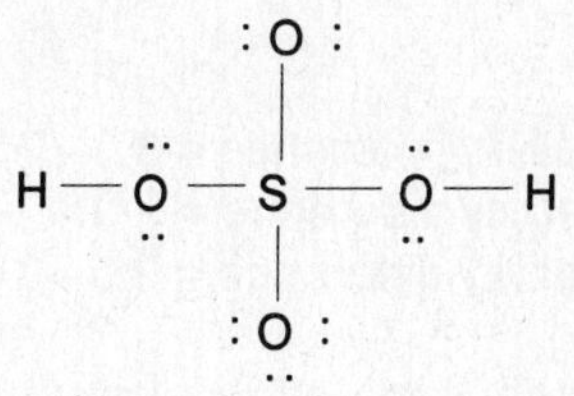 H—Ö—S(=O:)(=O:)—Ö—H

c. 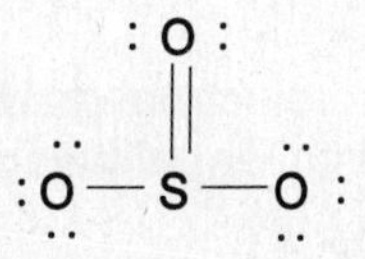 :Ö—S(=O:)—Ö:

d. 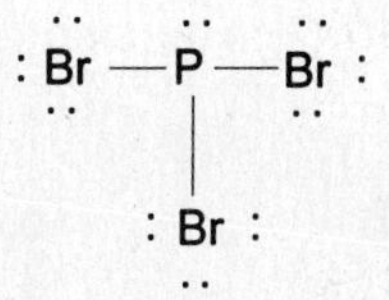 :B̈r—P(Br:)—B̈r:

11.51 a. In SO_2, 3 electron groups surround the central atom so the arrangement is trigonal planar.

b. In $SeCl_2$, 4 electron groups surround the central atom so the arrangement is tetrahedral.

c. In ClO_4^-, 4 electron groups surround the central atom so the arrangement is tetrahedral.

d. In PH_3, 4 electron groups surround the central atom so the arrangement is tetrahedral.

152 ■ CHAPTER 11

11.52 a. In $AsBr_3$, 4 electron groups surround the central atom so the arrangement is tetrahedral.

b. In HBrO, 4 electron groups surround the central atom so the arrangement is tetrahedral.

c. In ClCN, 2 electron groups surround the central atom so the arrangement is linear.

d. In Cl_2O, 4 electron groups surround the central atom so the arrangement is tetrahedral.

11.53 a. tetrahedral shape b. bent or angular shape
c. trigonal planar shape d. tetrahedral shape

11.54 a. tetrahedral shape b. tetrahedral shape
c. trigonal planar shape d. trigonal pyramidal shape

11.55 H–Cl electronegativity difference = 3.0 - 2.1 = 0.9
P–Cl electronegativity difference = 3.0 - 2.1 = 0.9
N–Cl electronegativity difference = 3.0 - 3.0 = 0.0

The smallest difference in electronegativity, and thus the least polar bond, is N–Cl.

11.56 (See problem #55 for electronegativity differences) Since the electronegativity difference for H-Cl and for P-Cl is the same, they have bonds of equal polarity.

11.57 a. ionic (metal + nonmetal) b. covalent (two nonmetals)
c. ionic (metal + nonmetal) d. covalent (two nonmetals)

11.58 a. ionic (metal + nonmetal) b. covalent (two nonmetals)
c. covalent (metalloid + nonmetal) d. ionic (metal + nonmetal)

11.59 a. HF b. NF_3 c. H_2O

11.60 a. HCl b. SO_2 c. PH_3

12. THE GASEOUS STATE

■ Solutions to Exercises

<u>Note on significant figures:</u> The first time the final answer is written, it is written first with one nonsignificant figure. The least significant digit is also underlined. The final answer is then rounded to the correct number of significant figures but the least significant digit is no longer underlined.

12.1 $0.712 \text{ atm} \times \dfrac{760 \text{ mm Hg}}{1 \text{ atm}} = 54\underline{1}.1 \text{ mm Hg} = 541 \text{ mm Hg}$

12.2 The quantities you need are

$P_1 = 1.00 \text{ atm}$
$V_1 = 20.0 \text{ L}$

$P_2 = 2.10 \text{ atm}$
$V_2 = ?$

Notice that the temperature, 23°C, is the same in both conditions, so it is constant. Thus you know

$$V_2 = V_1 \times \text{pressure ratio}$$

The volume must decrease since the pressure increases. Thus the pressure ratio must be <u>less</u> than one. This gives

$$V_2 = 20.0 \text{ L} \times \dfrac{1.00 \text{ atm}}{2.10 \text{ atm}} = 9.5\underline{2}4 \text{ L} = 9.52 \text{ L}$$

154 ■ CHAPTER 12

12.3 The quantities you need are

P_1 = 745 mm Hg
V_1 = 1.32 L

→

P_2 = ?
V_2 = 1.89 L

Notice that the temperature, 18°C, is not changing, so it is constant. The equation you want is

$P_2 = P_1 \times$ volume ratio

Since the volume increases, the pressure must decrease. Thus the volume ratio must be <u>less</u> than one. This gives

P_2 = 745 mm Hg × $\dfrac{1.32 \text{ L}}{1.89 \text{ L}}$ = 520.3 mm Hg = 5.20 × 10² mm Hg

12.4 Convert the temperatures to the Kelvin scale:

initial temperature (T_1) = 19 + 273 = 292 K

final temperature (T_2) = 25 + 273 = 298 K

The quantities you need for this problem are

V_1 = 438 mL
T_1 = 292 K

→

V_2 = ?
T_2 = 298 K

Notice that there is no change in pressure, 742 mm Hg, so it is constant. The equation you want is

$V_2 = V_1 \times$ temperature ratio

Since the temperature increases, the volume must also increase. Thus, the temperature ratio must be <u>greater</u> than one. This gives

V_2 = 438 mL × $\dfrac{298 \text{ K}}{292 \text{ K}}$ = 447.0 mL = 447 mL

THE GASEOUS STATE ■ 155

12.5 First, convert the initial temperature to the Kelvin scale

initial temperature (T_1) = 50.0 + 273 = 323 K

Next, identify the quantities you need:

V_1 = 10.0 mL
T_1 = 323 K

→

V_2 = 20.0 mL
T_2 = ?

Notice that the pressure, 736 mm Hg, is not changing so is constant. The equation you want is

$T_2 = T_1$ x volume ratio

Since the volume increases, the temperature must also increase, so the volume ratio must be <u>greater</u> than one. This gives

$$T_2 = 323 \text{ K} \times \frac{20.0 \text{ mL}}{10.0 \text{ mL}} = 646.0 \text{ K}$$

Finally, convert the final temperature to the Celsius scale.

$T_{°C} = T_K - 273 = 646.0 - 273 = 37\underline{3}.0°C = 373°C$

12.6 You must first calculate how many moles of gas are in 3.0 L at STP. Since one mole occupies a volume of 22.4 L at STP, you get

$$3.0 \text{ L} \times \frac{1 \text{ mol } CO_2}{22.4 \text{ L}} = 0.134 \text{ mol } CO_2$$

Since 5.9 g of CO_2 is equivalent to 0.134 mol, the molar mass is

$$\frac{5.9 \text{ g}}{0.134 \text{ mol}} = 4\underline{4}.0 \text{ g/mol} = 44 \text{ g/mol}$$

12.7 You must find the volume of chlorine at STP. First, convert the temperatures to kelvins.

$$\text{initial temperature } (T_1) = 255 + 273 = 528 \text{ K}$$

$$\text{final temperature } (T_2) = 0 + 273 = 273 \text{ K}$$

The equation you need is

$$V_2 = V_1 \times \text{pressure ratio} \times \text{temperature ratio}$$

The quantities you need for this problem are

P_1 = 5.11 atm		P_2 = 1.00 atm
V_1 = 119 mL	→	V_2 = ?
T_1 = 528 K		T_2 = 273 K

Since the pressure decreases, the volume will increase so the pressure ratio is <u>greater</u> than one. Since the temperature decreases, the volume will also decrease. Thus the temperature ratio is <u>less</u> than one. This gives

$$V_2 = 119 \text{ mL} \times \frac{5.11 \text{ atm}}{1.00 \text{ atm}} \times \frac{273 \text{ K}}{528 \text{ K}} = 314.4 \text{ mL}$$

The volume must be expressed in liters. This gives

$$314.4 \text{ mL} \times \frac{1 \text{ L}}{1000 \text{ mL}} = 0.3144 \text{ L}$$

The molar mass of chlorine can now be calculated using the molar volume. Since 1.00 g of chlorine is present, you get

$$\frac{1.00 \text{ g}}{0.3144 \text{ L}} \times \frac{22.4 \text{ L}}{1 \text{ mol}} = 71.246 \text{ g/mol} = 71.2 \text{ g/mol}$$

THE GASEOUS STATE ■ 157

12.8 First, change the temperature to the Kelvin scale.

$$T_K = 17 + 273 = 290 \text{ K}$$

The quantities that you need to use in the calculation are

P = ?	n = 0.500 moles
V = 3.06 L	R = 0.08206 L atm / (K mol)
	T = 290 K

The pressure can now be calculated.

$$P = \frac{nRT}{V} = \frac{0.500 \text{ mol} \times 0.08206 \text{ L atm}/(\text{K mol}) \times 290 \text{ K}}{3.06 \text{ L}}$$

$$= 3.888 \text{ atm} = 3.89 \text{ atm}$$

12.9 First, change the temperature to kelvins.

$$T_K = 5 + 273 = 278 \text{ K}$$

The quantities that you need to use in the calculation are

P = 5.00 atm	n = 5.00 moles
V = ?	R = 0.08206 L atm / (K mol)
	T = 278 K

The volume can now be calculated.

$$V = \frac{nRT}{P} = \frac{5.00 \text{ mol} \times 0.08206 \text{ L atm}/(\text{K mol}) \times 278 \text{ K}}{5.00 \text{ atm}} = 22.81 \text{ L} = 22.8 \text{ L}$$

158 ■ CHAPTER 12

12.10 The quantities used in the calculation are

P = 0.886 atm	n = ?
V = 0.452 L	R = 0.08206 L atm / (K mol)
	T = 298 K

The moles of gas can now be calculated.

$$n = \frac{PV}{RT} = \frac{0.886 \text{ atm} \times 0.452 \text{ L}}{0.08206 \text{ L atm}/(\text{K mol}) \times 298 \text{ K}} = 0.01638 \text{ mol} = 0.0164 \text{ mol}$$

12.11 First, convert the pressure and volume to atmospheres and liters, respectively, so that they match the units for the constant, R.

$$V = 894 \text{ mL} \times \frac{1 \text{ L}}{1000 \text{ mL}} = 0.894 \text{ L}$$

$$P = 753 \text{ mm Hg} \times \frac{1 \text{ atm}}{760 \text{ mm Hg}} = 0.9908 \text{ atm}$$

The quantities that you need for the calculation are

P = 0.9908 atm	n = 0.134 mol
V = 0.894 L	R = 0.08206 L atm / (K mol)
	T = ?

The equation that you need is

$$T = \frac{PV}{nR} = \frac{0.9908 \text{ atm} \times 0.894 \text{ L}}{0.134 \text{ mol} \times 0.08206 \text{ L atm}/(\text{K mol})} = 80.55 \text{ K}$$

Finally, convert the temperature to Celsius.

$$T_{°C} = 80.55 - 273 = -192.4 \text{ °C} = -192 \text{ °C}$$

12.12 First, convert the temperature to kelvins.

$$T_K = 25 + 273 = 298 \text{ K}$$

Next, calculate the pressure of each component of the mixture in the flask. The quantities that you need for carbon dioxide are

```
P = ?                    n = 0.0200 moles
V = 2.00 L               R = 0.08206 L atm/(K mol)
                         T = 298 K
```

Substitution leads to

$$P = \frac{nRT}{V} = \frac{0.0200 \text{ mol} \times 0.08206 \text{ L atm/(K mol)} \times 298 \text{ K}}{2.00 \text{ L}} = 0.2445 \text{ atm} = 0.245 \text{ atm}$$

The quantities that you need for oxygen are

```
P = ?                    n = 0.0400 moles
V = 2.00 L               R = 0.08206 L atm/(K mol)
                         T = 298 K
```

Substitution leads to

$$P = \frac{nRT}{V} = \frac{0.0400 \text{ mol} \times 0.08206 \text{ L atm/(K mol)} \times 298 \text{ K}}{2.00 \text{ L}} = 0.4891 \text{ atm} = 0.489 \text{ atm}$$

The total pressure in the flask is

$$P_T = P_{CO_2} + P_{O_2} = 0.2445 \text{ atm} + 0.4891 \text{ atm} = 0.7336 \text{ atm} = 0.734 \text{ atm}$$

CHAPTER 12

12.13 The vapor pressure of water at 17°C is 14.5 mm Hg. This pressure must be converted to atmospheres.

$$P = 14.5 \text{ mm Hg} \times \frac{1 \text{ atm}}{760 \text{ mm Hg}} = 0.01908 \text{ atm}$$

The volume is 203 mL, or 0.203 L, and the temperature is 17°C + 273 = 290 K. The quantities that you need are

$P = 0.01908$ atm $\qquad n = ?$

$V = 0.203$ L $\qquad R = 0.08206$ L atm/ (K mol)

$\qquad\qquad\qquad\qquad T = 290$ K

Substitution leads to

$$n = \frac{PV}{RT} = \frac{0.01908 \text{ atm} \times 0.203 \text{ L}}{0.08206 \text{ L atm/ (K mol)} \times 290 \text{ K}} = 1.628 \times 10^{-4} \text{ mol} = 1.63 \times 10^{-4} \text{ mol}$$

12.14 First, you must calculate the number of moles of calcium carbonate. The molar mass of $CaCO_3$ is 100.09 g/mol. This gives

$$\text{moles } CaCO_3 = 15.0 \text{ g } CaCO_3 \times \frac{1 \text{ mol } CaCO_3}{100.09 \text{ g } CaCO_3} = 0.1499 \text{ mol } CaCO_3$$

Second, you must calculate the number of moles of carbon dioxide by recognizing that each mole of calcium carbonate leads to one mole of carbon dioxide, or

$$\text{moles of } CO_2 = 0.1499 \text{ mol } CaCO_3 \times \frac{1 \text{ mol } CO_2}{1 \text{ mol } CaCO_3} = 0.1499 \text{ mol } CO_2$$

For the final step, you need to convert the pressure and temperature to atmospheres and kelvins, respectively.

$$P = 745 \text{ mm Hg} \times \frac{1 \text{ atm}}{760 \text{ mm Hg}} = 0.9803 \text{ atm}$$

$T_K = 27 + 273 = 300$ K

Substitution leads to

$$V = \frac{nRT}{P} = \frac{0.1499 \text{ mol} \times 0.08206 \text{ L atm/ (K mol)} \times 300 \text{ K}}{0.9803 \text{ atm}} = 3.764 \text{ L} = 3.76 \text{ L}$$

Answers to Questions to Test Your Reading

12.1 The three states of matter are <u>gases</u>, <u>liquids</u>, and <u>solids</u>.

12.2 The four properties of all gases are
  i. Gases fill a container completely and uniformly.
  ii. Gases are compressible.
  iii. Gases have low densities.
  iv. Gases exert a uniform pressure on all inner surfaces of a container because it completely fills the container.

12.3 <u>Pressure</u> is force exerted on a unit area. Atmospheric pressure is measured with a barometer. The height of the column of mercury in the barometer, in mm, is equal to the atmospheric pressure, in mm Hg.

12.4 The atmosphere around the earth contains about 5×10^{18} kg of gaseous substances, which exerts a force on every part of the earth's surface because of the earth's gravitational attraction. This force per unit area is atmospheric pressure.

12.5 The average atmospheric pressure at sea level is 760 mm Hg.

12.6 A <u>manometer</u> is a device that measures a gas's pressure in a container. The manometer is an open ended U-tube containing mercury (see figure 12.5). The difference in the heights of mercury in the two arms of the U-tube indicates the pressure difference, in mm Hg, between the gas in the container and the atmospheric pressure.

12.7 The pressure of a gas inside a perfectly elastic balloon is equal to the atmospheric pressure outside. Thus, the pressure inside the balloon would be 1 atm.

12.8 According to the <u>kinetic molecular theory of gases</u>, gases exert a pressure on any surface that they touch because the gas particles collide with the surface and push against it.

12.9 <u>Boyle's law</u> states that the volume of a fixed molar amount of gas at a given temperature is inversely proportional to the applied pressure.

12.10 Since, according to Boyle's law, at constant temperature pressure and volume are inversely related. Thus, the lower the pressure, the larger the volume. Since the atmospheric pressure is lower at higher altitudes, the volume would be larger in Denver, Colorado.

12.11 The fourth postulate of the theory states that the average kinetic energy of gas particles is directly proportional to the Kelvin temperature. As a result, the average energy of a collision when one of the gas particles hits a wall of the container will not change when the volume decreases at constant temperature. However, the pressure will <u>increase</u> when the volume <u>decreases</u> because of more collisions with the container per unit area. Thus, pressure and volume should be <u>inversely</u> related as Boyle's law states.

12.12 <u>Charles' law</u> states that the volume of a fixed amount of gas is directly proportional to its temperature as long as the pressure is kept constant.

12.13 Since, according to Charles' law, at constant pressure the volume and temperature are directly related. Thus, the higher the temperature, the larger the volume. Therefore, the balloon would be larger on a hot summer day.

12.14 Since the pressure of a gas inside a non-rigid container must always equal the atmospheric pressure, the pressure of the gas will not change when the temperature changes. For a decrease in temperature, the average kinetic energy of the gas particles will decrease. Less powerful collisions with the walls of the container will occur and it will contract. Thus, when the temperature <u>decreases</u>, the volume will <u>decrease</u>. In accord with Charles' law, volume and temperature are <u>directly</u> proportional.

12.15 STP, or <u>standard temperature and pressure</u>, is an arbitrarily chosen reference state equal to a temperature of 0°C and a pressure of 1 atmosphere.

12.16 <u>Avogadro's law</u> can be stated in two different ways:
i. Equal volumes of any two gases at the same temperature and pressure contain equal moles of molecules, and
ii. The volume of a gas is directly proportional to the moles of the gas as long as the temperature and pressure are constant.

12.17 Since the dimensions of gas particles are very small compared to the very large average distance between the particles, the sizes of the gas particles is not a factor. Gases are mostly empty space.

12.18 The <u>ideal gas law</u> can be written as $PV = nRT$, where R is the molar gas constant.

12.19 <u>Dalton's law of partial pressures</u> states that the total pressure exerted by a gaseous mixture is the sum of the partial pressures of the components of the mixture. The partial pressure of a gas in a mixture is the pressure the gas would exert if it were in the container alone.

12.20 The <u>vapor pressure</u> of water (or another liquid) is the pressure that builds up over a liquid in a closed container due to evaporation. For gases collected over water, it is the partial pressure of the water vapor. The vapor pressure of water at 20°C is equal to 17.5 mm Hg (see Table 12.2).

■ Solutions to Practice Problems

Note on significant figures: The first time the final answer is written, it is written first with one nonsignificant figure. The least significant digit is also underlined. The final answer is then rounded to the correct number of significant figures but the least significant digit is no longer underlined.

12.21 1128 mm Hg × $\dfrac{1 \text{ atm}}{760 \text{ mm Hg}}$ = 1.48̲4 atm = 1.48 atm

12.22 0.912 atm × $\dfrac{760 \text{ mm Hg}}{1 \text{ atm}}$ = 693̲.1 mm Hg = 693 mm Hg

12.23 V_2 = V_1 × pressure ratio = 2.60 L × $\dfrac{768 \text{ mm Hg}}{614 \text{ mm Hg}}$ = 3.25̲2 L = 3.25 L

12.24 V_2 = V_1 × pressure ratio = 6.50 L × $\dfrac{1.50 \text{ atm}}{2.50 \text{ atm}}$ = 3.90̲0 L = 3.90 L

12.25 V_2 = V_1 × pressure ratio = 849 L × $\dfrac{1.00 \text{ atm}}{21.6 \text{ atm}}$ = 39.3̲1 L = 39.3 L

12.26 V_2 = V_1 × pressure ratio = 24.8 L × $\dfrac{1.05 \text{ atm}}{6.50 \text{ atm}}$ = 4.00̲6 L = 4.01 L

12.27 P_2 = P_1 × volume ratio = 763 mm Hg × $\dfrac{845 \text{ mL}}{643 \text{ mL}}$

= 100̲3 mm Hg = 1.00 × 10^3 mm Hg

12.28 P_2 = P_1 × volume ratio = 1.90 atm × $\dfrac{1.05 \text{ L}}{3.45 \text{ L}}$ = 0.578̲3 atm = 0.578 atm

12.29 P_2 = P_1 × volume ratio = 1.00 atm × $\dfrac{145 \text{ L}}{165 \text{ L}}$ = 0.878̲8 atm = 0.879 atm

12.30 P_2 = P_1 × volume ratio = 3.08 atm × $\dfrac{208 \text{ mL}}{345 \text{ mL}}$ = 1.85̲7 atm = 1.86 atm

12.31 initial temperature (T_1) = 25 + 273 = 298 K
final temperature (T_2) = 0 + 273 = 273 K

$$V_2 = V_1 \times \text{temperature ratio} = 1.33 \times 10^3 \text{ L} \times \frac{273 \text{ K}}{298 \text{ K}}$$

$$= 12\underline{1}8 \text{ L} = 1.22 \times 10^3 \text{ L}$$

12.32 initial temperature (T_1) = 31 + 273 = 304 K
final temperature (T_2) = 0 + 273 = 273 K

$$V_2 = V_1 \times \text{temperature ratio} = 41.3 \text{ mL} \times \frac{273 \text{ K}}{304 \text{ K}}$$

$$= 37.\underline{0}9 \text{ mL} = 37.1 \text{ mL}$$

12.33 initial temperature (T_1) = 26 + 273 = 299 K
final temperature (T_2) = 14 + 273 = 287 K

$$V_2 = V_1 \times \text{temperature ratio} = 125 \text{ mL} \times \frac{287 \text{ K}}{299 \text{ K}}$$

$$= 12\underline{0}.0 \text{ mL} = 1.20 \times 10^2 \text{ mL}$$

12.34 initial temperature (T_1) = 51 + 273 = 324 K
final temperature (T_2) = -111 + 273 = 162 K

$$V_2 = V_1 \times \text{temperature ratio} = 222 \text{ mL} \times \frac{162 \text{ K}}{324 \text{ K}}$$

$$= 11\underline{1}.0 \text{ mL} = 111 \text{ mL}$$

12.35 initial temperature (T_1) = 0 + 273 = 273 K

$$T_2 = T_1 \times \text{volume ratio} = 273 \text{ K} \times \frac{3.22 \text{ L}}{2.22 \text{ L}} = 396.0 \text{ K}$$

final temperature (T_2) = 396.0 - 273 = 12\underline{3}.0°C = 123°C

12.36 initial temperature $(T_1) = 0 + 273 = 273$ K

$T_2 = T_1 \times$ volume ratio $= 273$ K $\times \dfrac{1.22 \text{ L}}{2.22 \text{ L}} = 150.0$ K

final temperature $(T_2) = 150.0 - 273 = -123.0°C = -123°C$

12.37 initial temperature $(T_1) = 30 + 273 = 303$ K

$T_2 = T_1 \times$ volume ratio $= 303$ K $\times \dfrac{4.00 \text{ L}}{2.00 \text{ L}} = 606.0$ K

final temperature $(T_2) = 606.0 - 273 = 333.0°C = 333°C$

12.38 initial temperature $(T_1) = 30.0 + 273 = 303.0$ K

$T_2 = T_1 \times$ volume ratio $= 303$ K $\times \dfrac{1.00 \text{ L}}{2.00 \text{ L}} = 151.5$ K

final temperature $(T_2) = 151.5 - 273 = -121.5°C = -122°C$

12.39 $\dfrac{1.00 \text{ g}}{11.1 \text{ L}} \times \dfrac{22.4 \text{ L}}{1 \text{ mol}} = 2.018$ g/mol $= 2.02$ g/mol

12.40 $\dfrac{1.00 \text{ g}}{0.700 \text{ L}} \times \dfrac{22.4 \text{ L}}{1 \text{ mol}} = 32.00$ g/mol $= 32.0$ g/mol

12.41 First, calculate the moles of argon. The temperature is 25°C + 273 = 298 K.

$n = \dfrac{PV}{RT} = \dfrac{2.00 \text{ atm} \times 0.561 \text{ L}}{0.08206 \text{ L atm/(K mol)} \times 298 \text{ K}} = 0.04588$ mol

The molar mass is therefore

$\dfrac{1.83 \text{ g}}{0.04588 \text{ mol}} = 39.89$ g/mol $= 39.9$ g/mol

12.42 First, calculate the moles of gas. The temperature is 25°C + 273 = 298 K.

$$n = \frac{PV}{RT} = \frac{1.56 \text{ atm} \times 1.74 \text{ L}}{0.08206 \text{ L atm}/(K \text{ mol}) \times 298 \text{ K}} = 0.1110 \text{ mol}$$

The molar mass is therefore

$$\frac{3.54 \text{ g}}{0.1110 \text{ mol}} = 31.89 \text{ g/mol} = 31.9 \text{ g/mol}$$

12.43 $T_K = 23 + 273 = 296$ K

$$P = \frac{nRT}{V} = \frac{3.03 \text{ mol} \times 0.08206 \text{ L atm}/(K \text{ mol}) \times 296 \text{ K}}{9.65 \text{ L}}$$

$$= 7.627 \text{ atm} = 7.63 \text{ atm}$$

12.44 $T_K = 19 + 273 = 292$ K

$$P = \frac{nRT}{V} = \frac{0.200 \text{ mol} \times 0.08206 \text{ L atm}/(K \text{ mol}) \times 292 \text{ K}}{6.00 \text{ L}}$$

$$= 0.7987 \text{ atm} = 0.799 \text{ atm}$$

12.45 $T_K = 100 + 273 = 373$ K

$$P = \frac{nRT}{V} = \frac{0.200 \text{ mol} \times 0.08206 \text{ L atm}/(K \text{ mol}) \times 373 \text{ K}}{5.00 \text{ L}}$$

$$= 1.224 \text{ atm} = 1.22 \text{ atm}$$

12.46 $T_K = 24 + 273 = 297$ K

$$P = \frac{nRT}{V} = \frac{3.00 \text{ mol} \times 0.08206 \text{ L atm}/(K \text{ mol}) \times 297 \text{ K}}{8.45 \text{ L}}$$

$$= 8.653 \text{ atm} = 8.65 \text{ atm}$$

12.47 $T_K = 55 + 273 = 328$ K

$$V = \frac{nRT}{P} = \frac{2.50 \text{ mol} \times 0.08206 \text{ L atm}/(K \text{ mol}) \times 328 \text{ K}}{1.50 \text{ atm}} = 44.86 \text{ L} = 44.9 \text{ L}$$

12.48 $T_K = 26 + 273 = 299$ K $P = 738 \text{ mm Hg} \times \dfrac{1 \text{ atm}}{760 \text{ mm Hg}} = 0.9711$ atm

$$V = \dfrac{nRT}{P} = \dfrac{0.120 \text{ mol} \times 0.08206 \text{ L atm/(K mol)} \times 299 \text{ K}}{0.9711 \text{ atm}}$$

$$= 3.032 \text{ L} = 3.03 \text{ L}$$

12.49 $T_K = 75 + 273 = 348$ K $P = 775 \text{ mm Hg} \times \dfrac{1 \text{ atm}}{760 \text{ mm Hg}} = 1.020$ atm

$$V = \dfrac{nRT}{P} = \dfrac{0.125 \text{ mol} \times 0.08206 \text{ L atm/(K mol)} \times 348 \text{ K}}{1.020 \text{ atm}}$$

$$= 3.500 \text{ L} = 3.50 \text{ L}$$

12.50 $T_K = 22 + 273 = 295$ K

$$V = \dfrac{nRT}{P} = \dfrac{0.330 \text{ mol} \times 0.08206 \text{ L atm/(K mol)} \times 295 \text{ K}}{2.00 \text{ atm}} = 3.994 \text{ L} = 3.99 \text{ L}$$

12.51 $T_K = 26 + 273 = 299$ K $V = 125 \text{ mL} \times \dfrac{1 \text{ L}}{1000 \text{ mL}} = 0.125$ L

$P = 745 \text{ mm Hg} \times \dfrac{1 \text{ atm}}{760 \text{ mm Hg}} = 0.9803$ atm

$$n = \dfrac{PV}{RT} = \dfrac{0.9803 \text{ atm} \times 0.125 \text{ L}}{0.08206 \text{ L atm/(K mol)} \times 299 \text{ K}} = 0.004994 \text{ mol} = 4.99 \times 10^{-3} \text{ mol}$$

12.52 $T_K = 600 + 273 = 873$ K

$$n = \dfrac{PV}{RT} = \dfrac{1.12 \text{ atm} \times 2.68 \text{ L}}{0.08206 \text{ L atm/(K mol)} \times 873 \text{ K}} = 0.04189 \text{ mol} = 0.0419 \text{ mol}$$

12.53 $T_K = 20 + 273 = 293$ K $\quad V = 545 \text{ mL} \times \dfrac{1 \text{ L}}{1000 \text{ mL}} = 0.545$ L

$P = 760 \text{ mm Hg} \times \dfrac{1 \text{ atm}}{760 \text{ mm Hg}} = 1.00$ atm

$n = \dfrac{PV}{RT} = \dfrac{1.00 \text{ atm} \times 0.545 \text{ L}}{0.08206 \text{ L atm}/(\text{K mol}) \times 293 \text{ K}} = 0.022\underline{7} \text{ mol} = 0.023$ mol

12.54 This problem can be solved by two different methods. First, use the molar volume at STP:

moles $= 2.00 \text{ L} \times \dfrac{1 \text{ mol}}{22.4 \text{ L}} = 0.089\underline{2}9 \text{ mol} = 0.0893$ mol

Second, use the ideal gas law: $T_K = 0 + 273 = 273$ K

$n = \dfrac{PV}{RT} = \dfrac{1.000 \text{ atm} \times 2.00 \text{ L}}{0.08206 \text{ L atm}/(\text{K mol}) \times 273 \text{ K}} = 0.089\underline{2}7 \text{ mol} = 0.0893$ mol

12.55 $T = \dfrac{PV}{nR} = \dfrac{3.50 \text{ atm} \times 4.00 \text{ L}}{0.410 \text{ mol} \times 0.08206 \text{ L atm}/(\text{K mol})} = 416.1$ K

$T_{°C} = 416.1 - 273 = 14\underline{3}.1°C = 143°C$

12.56 $P = 956 \text{ mm Hg} \times \dfrac{1 \text{ atm}}{760 \text{ mm Hg}} = 1.258$ atm

$T = \dfrac{PV}{nR} = \dfrac{1.258 \text{ atm} \times 2.50 \text{ L}}{0.128 \text{ mol} \times 0.08206 \text{ L atm}/(\text{K mol})} = 299.4$ K

$T_{°C} = 299.4 - 273 = 2\underline{6}.4°C = 26°C$

12.57 $P = 755 \text{ mm Hg} \times \dfrac{1 \text{ atm}}{760 \text{ mm Hg}} = 0.9934$ atm

$T = \dfrac{PV}{nR} = \dfrac{0.9934 \text{ atm} \times 10.0 \text{ L}}{1.00 \text{ mol} \times 0.08206 \text{ L atm}/(\text{K mol})} = 121.1$ K

$T_{°C} = 121.1 - 273 = -15\underline{1}.9°C = -152°C$

12.58 $V = 253 \text{ mL} \times \dfrac{1 \text{ L}}{1000 \text{ mL}} = 0.253 \text{ L}$

$T = \dfrac{PV}{nR} = \dfrac{1.11 \text{ atm} \times 0.253 \text{ L}}{0.250 \text{ mol} \times 0.08206 \text{ L atm}/(\text{K mol})} = 13.69 \text{ K}$

$T_{°C} = 13.69 - 273 = -25\underline{9}.3\text{°C} = -259\text{°C}$

12.59 $T_K = 15 + 273 = 288 \text{ K}$

$P_{O_2} = \dfrac{nRT}{V} = \dfrac{0.00125 \text{ mol} \times 0.08206 \text{ L atm}/(\text{K mol}) \times 288 \text{ K}}{2.00 \text{ L}} = 0.01477 \text{ atm}$

$P_{He} = \dfrac{nRT}{V} = \dfrac{0.00325 \text{ mol} \times 0.08206 \text{ L atm}/(\text{K mol}) \times 288 \text{ K}}{2.00 \text{ L}} = 0.03840 \text{ atm}$

$P_{Tot} = P_{O_2} + P_{He} = 0.01477 \text{ atm} + 0.03840 \text{ atm} = 0.053\underline{1}7 \text{ atm} = 0.0532 \text{ atm}$

12.60 $T_K = 15 + 273 = 288 \text{ K}$. The partial pressures are

$P_{N_2} = \dfrac{n_{N_2}RT}{V} = \dfrac{1.00 \text{ mol} \times 0.08206 \text{ L atm}/(\text{K mol}) \times 288 \text{ K}}{5.00 \text{ L}} = 4.7\underline{2}7 \text{ atm} = 4.73 \text{ atm}$

$P_{H_2} = \dfrac{n_{H_2}RT}{V} = \dfrac{2.00 \text{ mol} \times 0.08206 \text{ L atm}/(\text{K mol}) \times 288 \text{ K}}{5.00 \text{ L}} = 9.4\underline{5}3 \text{ atm} = 9.45 \text{ atm}$

The total pressure is

$P_{tot} = P_{N_2} + P_{H_2} = 4.727 \text{ atm} + 9.453 \text{ atm} = 14.1\underline{8}0 = 14.18 \text{ atm}$

170 ■ CHAPTER 12

12.61 The vapor pressure of water at 30 °C is 31.8 mm Hg (see Table 12.2). The partial pressure of hydrogen can now be determined.

$$P_{H_2} = P_{Tot} - P_{H_2O} = 752 \text{ mm Hg} - 31.8 \text{ mm Hg} = 720.2 \text{ mm Hg}$$

The ideal gas law can now be used to find the number of moles.

$$T_K = 30 + 273 = 303 \text{ K} \qquad V = 232 \text{ mL} \times \frac{1 \text{ L}}{1000 \text{ mL}} = 0.232 \text{ L}$$

$$P = 720.2 \text{ mm Hg} \times \frac{1 \text{ atm}}{760 \text{ mm Hg}} = 0.9476 \text{ atm}$$

$$n = \frac{PV}{RT} = \frac{0.9476 \text{ atm} \times 0.232 \text{ L}}{0.08206 \text{ L atm}/(K \text{ mol}) \times 303 \text{ K}} = 0.00884\underline{2} \text{ mol} = 8.84 \times 10^{-3} \text{ mol}$$

12.62 The vapor pressure of water at 45°C is 73.9 mm Hg (see Table 12.2). The pressure must be converted to atmospheres.

$$P = 73.9 \text{ mm Hg} \times \frac{1 \text{ atm}}{760 \text{ mm Hg}} = 0.09724 \text{ atm}$$

The pressure of argon can now be determined.

$$P_{Ar} = P_{Tot} - P_{H_2O} = 1.05 \text{ atm} - 0.09724 \text{ atm} = 0.9528 \text{ atm}$$

The deal gas law can now be used to find the moles of argon.

$$T_K = 45 + 273 = 318 \text{ K}$$

$$n = \frac{PV}{RT} = \frac{0.9528 \text{ atm} \times 4.52 \text{ L}}{0.08206 \text{ L atm}/(K \text{ mol}) \times 318 \text{ K}} = 0.165\underline{0} \text{ mol} = 0.165 \text{ mol}$$

12.63 The first step is to calculate the number of moles of methane gas. The molar mass of CH_4 is 16.04 g/mol.

$$\text{moles } CH_4 = 10.0 \text{ g} \times \frac{1 \text{ mol}}{16.04 \text{ g}} = 0.6234 \text{ mol } CH_4$$

THE GASEOUS STATE ■ 171

The second step is to calculate the number of moles of oxygen by recognizing that each mole of carbon dioxide requires two moles of oxygen.

$$\text{moles } O_2 = 0.6234 \text{ mol CH}_4 \times \frac{2 \text{ mol } O_2}{1 \text{ mol CH}_4} = 1.248 \text{ mol } O_2$$

The final step is to use the ideal gas equation to calculate the volume.

$$V = \frac{nRT}{P} = \frac{1.248 \text{ mol} \times 0.08206 \text{ L atm}/(K \text{ mol}) \times 273 \text{ K}}{1.00 \text{ atm}}$$

$$= 27.96 \text{ L} = 28.0 \text{ L}$$

12.64 The first step is to calculate the number of moles of ethane gas. The molar mass of C_2H_6 is 30.07 g/mol.

$$\text{moles } C_2H_6 = 10.0 \text{ g} \times \frac{1 \text{ mol } C_2H_6}{30.07 \text{ g}} = 0.3326 \text{ mol } C_2H_6$$

The second step is to calculate the number of moles of oxygen by recognizing that two moles of ethane require seven moles of oxygen.

$$\text{moles } O_2 = 0.3326 \text{ mol } C_2H_6 \times \frac{7 \text{ mol } O_2}{2 \text{ mol } C_2H_6} = 1.164 \text{ mol } O_2$$

The final step is to use the ideal equation to calculate the volume.

$$V = \frac{nRT}{P} = \frac{1.164 \text{ mol} \times 0.08206 \text{ L atm}/(K \text{ mol}) \times 273 \text{ K}}{1.00 \text{ atm}} = 26.07 \text{ L} = 26.1 \text{ L}$$

12.65 The first step is to calculate the number of moles of hydrogen peroxide. The molar mass of H_2O_2 is 34.02 g/mol.

$$\text{moles } H_2O_2 = 1.00 \text{ g} \times \frac{1 \text{ mol } H_2O_2}{34.02 \text{ g}} = 0.02939 \text{ mol } H_2O_2$$

The second step is to calculate the number of moles of oxygen by recognizing that two moles of hydrogen peroxide produce one mole of oxygen.

$$\text{moles } O_2 = 0.02939 \text{ mol } H_2O_2 \times \frac{1 \text{ mol } O_2}{2 \text{ mol } H_2O_2} = 0.01470 \text{ mol } O_2$$

172 ■ CHAPTER 12

The final step is to use the ideal gas equation to calculate the volume.

$$V = \frac{nRT}{P} = \frac{0.01470 \text{ mol} \times 0.08206 \text{ L atm/(K mol)} \times 273 \text{ K}}{1.00 \text{ atm}}$$

$$= 0.3293 \text{ L} = 0.329 \text{ L}$$

12.66 The first step is to calculate the number of moles of mercury (II) oxide. The molar mass of HgO is 216.59 g/mol.

$$\text{moles HgO} = 2.00 \text{ g} \times \frac{1 \text{ mol HgO}}{216.59 \text{ g}} = 0.009234 \text{ mol HgO}$$

The second step is to calculate the number of moles of oxygen by recognizing that two moles of mercury (II) oxide produce one mole of oxygen.

$$\text{moles O}_2 = 0.009234 \text{ mol HgO} \times \frac{1 \text{ mol O}_2}{2 \text{ mol HgO}} = 0.004617 \text{ mol O}_2$$

The final step is to use the ideal gas equation to calculate the volume.

$$V = \frac{nRT}{P} = \frac{0.004617 \text{ mol} \times 0.08206 \text{ L atm/(K mol)} \times 273 \text{ K}}{1.00 \text{ atm}}$$

$$= 0.1034 \text{ L} = 0.103 \text{ L}$$

Solutions to Additional Problems

Note on significant figures: The first time the final answer is written, it is written first with one nonsignificant figure. The least significant digit is also underlined. The final answer is then rounded to the correct number of significant figures but the least significant digit is no longer underlined.

12.67 The number of moles of oxygen from Problem #65 is 0.01470 mol.

$T_K = 23 + 273 = 296$ K

$$V = \frac{nRT}{P} = \frac{0.01470 \text{ mol} \times 0.08206 \text{ L atm/(K mol)} \times 296 \text{ K}}{1.12 \text{ atm}}$$

$= 0.31\underline{8}8$ L $= 0.319$ L

12.68 The number of moles of oxygen from Problem #66 is 0.004617 mol

$T_K = 16 + 273 = 289$ K

$$V = \frac{nRT}{P} = \frac{0.004617 \text{ mol} \times 0.08206 \text{ L atm/(K mol)} \times 289 \text{ K}}{0.756 \text{ atm}}$$

$= 0.14\underline{4}8$ L $= 0.145$ L

12.69 You must first determine the partial pressure of the helium.

$P_{He} = P_{Tot} - P_{O_2} = 3.00$ atm $- 0.200$ atm $= 2.800$ atm

Next, use the ideal gas law to find the number of moles.

$T_K = 23 + 273 = 296$ K

$$n = \frac{PV}{RT} = \frac{2.800 \text{ atm} \times 1.00 \text{ L}}{0.08206 \text{ L atm/(K mol)} \times 296 \text{ K}} = 0.11\underline{5}3 \text{ mol} = 0.115 \text{ mol}$$

174 ■ CHAPTER 12

12.70 First, calculate the total number of moles.

$$n = n_{O_2} + n_{He} = 1.50 \text{ mol} + 2.00 \text{ mol} = 3.50 \text{ mol}$$

Next, use the ideal gas law to find the volume.

$$V = \frac{nRT}{P} = \frac{3.50 \text{ mol} \times 0.08206 \text{ L atm}/(\text{K mol}) \times 298 \text{ K}}{0.9934 \text{ atm}} = 86.16 \text{ L} = 86.2 \text{ L}$$

12.71 First you must calculate the number of moles of helium that was produced.

$$P = 765 \text{ mm Hg} \times \frac{1 \text{ atm}}{760 \text{ mm Hg}} = 1.007 \text{ atm}$$

$$V = 12.5 \text{ mL} \times \frac{1 \text{ L}}{1000 \text{ mL}} = 0.0125 \text{ L} \qquad T_K = 23 + 273 = 296 \text{ K}$$

$$n_{He} = \frac{PV}{RT} = \frac{1.007 \text{ atm} \times 0.0125 \text{ L}}{0.08206 \text{ L atm}/(\text{K mol}) \times 296 \text{ K}} = 5.182 \times 10^{-4} \text{ mol He}$$

Second, you can calculate the atoms of helium.

$$\text{atoms He} = 5.182 \times 10^{-4} \text{ mol He} \times \frac{6.022 \times 10^{23} \text{ atoms He}}{1 \text{ mol He}} = 3.121 \times 10^{20} \text{ atoms He}$$

Finally, you can calculate the number of atoms of the metal by recognizing that each metal atom results in one helium atom.

$$\text{metal atoms} = 3.121 \times 10^{20} \text{ atoms He} \times \frac{1 \text{ metal atom}}{1 \text{ atom He}}$$

$$= 3.121 \times 10^{20} \text{ metal atoms} = 3.12 \times 10^{20} \text{ atoms}$$

12.72 The first step is to calculate the number of moles of carbon dioxide.

$$T_K = 0 + 273 = 273 \text{ K}$$

$$n = \frac{PV}{RT} = \frac{1.00 \text{ atm} \times 5.8 \times 10^2 \text{ L}}{0.08206 \text{ L atm}/(\text{K mol}) \times 273 \text{ K}} = 25.9 \text{ mol } CO_2$$

The second step is to calculate the number of moles of lithium hydroxide by recognizing that one mole of carbon dioxide requires two moles of lithium hydroxide.

$$\text{moles LiOH} = 25.9 \text{ mol } CO_2 \times \frac{2 \text{ mol LiOH}}{1 \text{ mol } CO_2} = 51.8 \text{ mol LiOH}$$

The final step is to calculate the number of grams of lithium hydroxide. The molar mass of LiOH is 23.95 g/mol.

$$\text{grams LiOH} = 51.8 \text{ mol LiOH} \times \frac{23.95 \text{ g LiOH}}{1 \text{ mol LiOH}} = 1241 \text{ g} = 1.2 \times 10^3 \text{ g}$$

13. LIQUIDS, SOLIDS, AND ATTRACTIONS BETWEEN MOLECULES

■ Solutions to Exercises

Note on significant figures: The first time the final answer is written, it is written first with one nonsignificant figure. The least significant digit is also underlined. The final answer is then rounded to the correct number of significant figures but the least significant digit is no longer underlined.

13.1 Frost is the condensation of water vapor to ice without going through the liquid state.

13.2 First, determine the number of moles in 1.0 g of mercury (molar mass 200.59 g/mol)

$$1.0 \text{ g} \times \frac{1 \text{ mol}}{200.59 \text{ g}} = 0.00499 \text{ mol}$$

Next, calculate the heat that must be removed.

$$0.00499 \text{ mol} \times \frac{2.3 \text{ kJ}}{1 \text{ mol}} = 0.01\underline{1}48 \text{ kJ} = 0.011 \text{ kJ}$$

13.3 The number of moles in 1.0 g of mercury is 0.00499 mol (see Exercise 2). The heat required is

$$0.00499 \text{ mol} \times \frac{59.3 \text{ kJ}}{1 \text{ mol}} = 0.2\underline{9}6 \text{ kJ} = 0.30 \text{ kJ}$$

LIQUIDS, SOLIDS, AND ATTRACTIONS BETWEEN MOLECULES ■ 177

13.4 This is a three step process. The first step is to calculate the heat that must be removed to convert 5.0 g of gaseous mercury at 357°C to liquid mercury. The number of moles of mercury is (molar mass 200.59 g/mol)

$$5.0 \text{ g} \times \frac{1 \text{ mol}}{200.59 \text{ g}} = 0.0249 \text{ mol}$$

The heat removed in this step is (the molar heat of vaporization is 59.3 kJ/mol)

$$0.0249 \text{ mol} \times \frac{59.3 \text{ kJ}}{1 \text{ mol}} = 1.48 \text{ kJ}$$

The second step is to calculate the heat that must be removed to lower the temperature from 357°C to -39°C, a change in temperature of 396°C (the specific heat is 0.14 J/g°C).

$$5.0 \text{ g} \times \frac{0.14 \text{ J}}{1 \text{ g °C}} \times 396°C = 277 \text{ J} = 0.277 \text{ kJ}$$

The third step is to calculate the heat that must be removed to change the liquid mercury into solid mercury at -39°C (the molar heat of fusion is 2.3 kJ/mol).

$$0.0249 \text{ mol} \times \frac{2.3 \text{ kJ}}{1 \text{ mol}} = 0.0573 \text{ kJ}$$

The overall heat that must be removed is

$$1.48 \text{ kJ} + 0.277 \text{ kJ} + 0.0573 \text{ kJ} = 1.\underline{8}1 \text{ kJ} = 1.8 \text{ kJ}$$

13.5 Both hydrogen (H_2) and bromine (Br_2) are nonpolar molecules. Therefore, in the liquid state only dispersion forces are present. Hydrogen bromide (HBr) is polar. Thus, in addition to dispersion forces, there are also dipole-dipole forces present in the liquid state of HBr.

13.6 Dispersion forces increase with increasing molecular weight. The molecular weights of CH_4, C_2H_6, and C_3H_8 are 16 amu, 30 amu, and 44 amu, respectively. Each of the three molecules is nonpolar so there are no dipole-dipole forces present. There is also no hydrogen bonding in any of the molecules. Thus, the order of increasing attractive forces is the same as the molecular weight. We conclude that the vapor pressure is greater where the forces are less. Thus, in order of increasing vapor pressure, we can conclude $C_3H_8 < C_2H_6 < CH_4$.

13.7 Both methane (CH_4) and silane (SiH_4) are symmetrical and nonpolar. Thus, the only forces present are dispersion forces. Since the dispersion forces increase with increasing molecular weight, there are more forces in silane (molecular weight 32.12 amu) than in methane (molecular weight 16.04 amu). Also, vapor pressure decreases with increasing forces so the vapor pressure of silane is less than for methane. Thus, the boiling point of SiH_4 should be higher.

13.8 Zinc (Zn) is a metal; it is a metallic solid. Sodium bromide (NaBr) is an ionic solid. Methane (CH_4) is a molecular substance and thus a molecular solid.

13.9 Magnesium chloride ($MgCl_2$) is an ionic solid. Methanol (CH_3OH) is a molecular substance and is a molecular solid. Argon (Ar) is a noble gas and thus an atomic solid. Argon is a gas at room temperature and has a very low melting point. Methanol is a liquid at room temperature so its melting point is higher than for argon. Magnesium chloride is a solid at room temperature and, since it is an ionic compound, it has a high melting point. Thus, in order of increasing melting point, we can conclude
Ar < CH_3OH < $MgCl_2$.

■ Answers to Questions to Test Your Reading

13.1 A <u>gas</u> has neither a fixed volume nor a fixed shape. A <u>liquid</u> has a fixed volume but it does not have a specific shape. A <u>solid</u> has a fixed volume and a fixed shape.

13.2 A change of state occurs when a substances changes from a solid, liquid, or gas into a different state. An example is when ice melts; water in the solid state has changed to water in the liquid state. This is called melting, or fusion.

13.3 The temperature is the principal factor that determines the state of matter, even though pressure also can be a factor. For example, at 1 atm pressure, water exists as a solid below 0°C, a liquid above 0°C and below 100°C, and a gas above 100°C.

13.4 The different changes of state are as follows:

<u>Melting</u> (or fusion), which is the change of a solid to a liquid. The melting of ice is an example.

$$H_2O\,(s) \longrightarrow H_2O\,(l)$$

<u>Freezing</u>, which is the change of a liquid to a solid. The freezing of water is an example.

$$H_2O\,(l) \longrightarrow H_2O\,(s)$$

<u>Vaporization</u>, which is the change of a liquid into a gas. The vaporization of water is an example.

$$H_2O\,(l) \longrightarrow H_2O\,(g)$$

LIQUIDS, SOLIDS, AND ATTRACTIONS BETWEEN MOLECULES ■ 179

Condensation, which is the change of a gas to a liquid, or a gas to a solid. An example is the condensation of water.

$$H_2O\ (g) \longrightarrow H_2O\ (l)$$

Sublimation, which is the change of a solid to a gas. The sublimation of water is an example.

$$H_2O\ (s) \longrightarrow H_2O\ (g)$$

13.5 At 25°C, water (H_2O) and mercury (Hg) are liquids. Sodium chloride (NaCl) and quartz (SiO_2) are solids at 25°C.

13.6 The melting point of ice is 0°C, and at this temperature, ice and water can coexist indefinitely in equilibrium, as long as no more heat is added. The normal boiling point of water (at 760 mm Hg) is 100°C. Since 120°C is above the normal boiling point of water, liquid water and steam cannot coexist indefinitely at this temperature.

13.7 The constant temperature at which a solid can coexist with its liquid is the melting point of the solid, <u>and</u> the freezing point of the liquid. For a given substance these are always identical.

13.8 The <u>normal melting point</u> of a solid is the constant temperature, at 1 atm pressure, at which the solid changes to a liquid.

13.9 The <u>normal boiling point</u> of a liquid is the constant temperature, at 1 atm pressure, at which the liquid changes to a gas.

13.10 The normal freezing point of water is 0°C and the normal boiling point is 100°C.

13.11 The <u>molar heat of fusion</u> is the energy required to melt (fuse) one mole of a solid. The <u>molar heat of vaporization</u> is the energy required to vaporize one mole of a liquid.

13.12 18 g (1 mole) of steam at 100°C will release 40.7 kJ of heat in becoming 18 g of liquid water at 100°C. Thus, 40.7 kJ is available to melt more ice.

13.13 When a liquid evaporates, it undergoes a phase change from liquid to gas. This requires that energy (heat) be absorbed from the remaining liquid. Thus, by removing heat from the liquid, it becomes cooler. As far as our bodies are concerned, water evaporates from our skin and removes heat in the process thereby cooling our bodies. This is called perspiration.

13.14 The <u>vapor pressure</u> of a liquid is the partial pressure of a gas over the liquid at equilibrium. Vapor pressure depends on the temperature. As the temperature increases, the vapor pressure increases.

13.15 Water will boil at 50°C if the atmospheric pressure is equal to the vapor pressure of water at this temperature, which is 92.5 mm Hg.

13.16 Below the boiling point, a liquid evaporates principally from its surface. In a closed container, a small quantity will evaporate, and then evaporation will appear to stop. This is because condensation is also occurring. The rate of evaporation and the rate of condensation will eventually become equal. This is called equilibrium.

13.17 Water droplets are spherical because a sphere has the smallest area per unit volume, which reduces the surface tension.

13.18 Dispersion forces are the weak attractive forces resulting from small, instantaneous dipoles that occur because of the varying positions of the electrons during their motion about nuclei. At any one instant in a molecule or atom, the electrons may not be uniformly distributed about the nuclei. For a brief instant, one side will have a small positive charge and the other side will have a small negative charge. This is called an instantaneous dipole. This instantaneous dipole can cause additional instantaneous dipoles in other nearby molecules. The result is that the two instantaneous dipoles attract each other as a dispersion force.

Since dispersion forces occur in the liquid state of all molecules, both liquid nitrogen and liquid water will experience these forces. Because N_2 is nonpolar, dispersion forces are the only type of intermolecular force present, and N_2 is a gas at room temperature. However, H_2O is polar and other forces, including hydrogen bonding are present, so dispersion forces play a small role in the properties of water.

13.19 A dipole-dipole force is an attractive intermolecular force resulting from the interaction of the positive end of one molecule with the negative end of another. These forces are important in any molecule that is unsymmetric, and therefore polar. An example of a liquid in which these forces are important is H_2O.

13.20 Hydrogen bonding occurs between two polar molecules containing a hydrogen atom covalently bonded to an atom of either nitrogen, oxygen, or fluorine. An example of a liquid in which these forces are important is methanol, CH_3OH.

13.21 If water was a linear molecule, it would be symmetric, and thus nonpolar.

13.22 A crystalline solid is a solid that is composed of one or more crystals with each crystal having a well-defined, ordered arrangement of the structural units. Examples are sodium chloride and sucrose. An amorphous solid is a solid that lacks order, and as a result, it does not have a well-defined arrangement of the structural units. Glass is an example.

13.23 An atomic solid is a solid containing atoms from a single element. An example is neon. A metallic solid is a solid containing atoms from a metallic element, such as copper. A covalent network solid is a solid containing atoms held in large networks or chains by covalent bonds. An example is carbon. A molecular solid is a solid consisting of molecules. Water (ice) is an example. An ionic solid is a solid consisting of ions. Sodium chloride is an example.

LIQUIDS, SOLIDS, AND ATTRACTIONS BETWEEN MOLECULES ■ 181

13.24 Ionic solids are held together by strong attractive forces that exist between oppositely charged ions. These forces are much stronger than those that exist between the partial positive and negative charges found in molecular solids. These strong forces cause ionic solids to have high melting points. Atomic solids come in a variety of types. Some are held together by only dispersion forces and have very low melting points. Noble gases are of this type. Metallic solids, which are also atomic solids, are held together by stronger forces and have relatively high melting points. A third type of atomic solid is for certain nonmetals like carbon and silicon. These are covalently bonded solids and have high melting points.

13.25 Covalent network solids have relatively high melting points because the strong covalent bonds must be broken.

13.26 Molecular solids can have either relatively low or relatively high melting points because, in addition to dispersion forces, some compounds include dipole-dipole forces or hydrogen bonding (high melting points), and some contain only dispersion forces (low melting points).

■ Solutions to Practice Problems

<u>Note on significant figures:</u> The first time the final answer is written, it is written first with one nonsignificant figure. The least significant digit is also underlined. The final answer is then rounded to the correct number of significant figures but the least significant digit is no longer underlined.

13.27a. Chlorine gas liquefies. This is called <u>condensation</u>.

b. Molten lava becomes a solid. This is <u>freezing</u>, or <u>fusion</u>.

13.28a. Moth balls slowly disappear due to a phase change from solid to gas. This phenomenon is called <u>sublimation</u>.

b. Water is changing from liquid to solid, or <u>freezing</u>.

182 ■ CHAPTER 13

13.29 First, calculate the number of moles in 75.0 g of ice. The molar mass of ice is 18.02 g/mol.

$$75.0 \text{ g} \times \frac{1 \text{ mol}}{18.02 \text{ g}} = 4.162 \text{ mol}$$

Next, calculate the heat required by using the molar heat of fusion.

$$4.162 \text{ mol} \times \frac{6.01 \text{ kJ}}{1 \text{ mol}} = 25.01 \text{ kJ} = 25.0 \text{ kJ}$$

13.30 First, calculate the number of moles in 75.0 g of water, molar mass 18.02 g/mol.

$$75.0 \text{ g} \times \frac{1 \text{ mol}}{18.02 \text{ g}} = 4.162 \text{ mol}$$

Next, calculate the heat produced.

$$4.162 \text{ mol} \times \frac{6.01 \text{ kJ}}{1 \text{ mol}} = 25.01 \text{ kJ} = 25.0 \text{ kJ}$$

13.31 First, calculate the number of moles in 26.4 g of ammonia (NH_3). The molar mass of ammonia is 17.03 g/mol.

$$26.4 \text{ g} \times \frac{1 \text{ mol}}{17.03 \text{ g}} = 1.550 \text{ mol}$$

Next, calculate the heat required to melt the ammonia.

$$1.550 \text{ mol} \times \frac{5.65 \text{ kJ}}{1 \text{ mol}} = 8.758 \text{ kJ} = 8.76 \text{ kJ}$$

13.32 First, calculate the number of moles in 26.4 g of ammonia (NH_3), molar mass 17.03 g/mol.

$$26.4 \text{ g} \times \frac{1 \text{ mol}}{17.03 \text{ g}} = 1.550 \text{ mol}$$

Next, calculate the heat produced.

$$1.550 \text{ mol} \times \frac{5.65 \text{ kJ}}{1 \text{ mol}} = 8.758 \text{ kJ} = 8.76 \text{ kJ}$$

LIQUIDS, SOLIDS, AND ATTRACTIONS BETWEEN MOLECULES ■ 183

13.33 First, calculate the number of moles in 75.0 g of water, molar mass 18.02 g/mol.

$$75.0 \text{ g} \times \frac{1 \text{ mol}}{18.02 \text{ g}} = 4.162 \text{ mol}$$

Next, calculate the heat required.

$$4.162 \text{ mol} \times \frac{40.7 \text{ kJ}}{1 \text{ mol}} = 169.4 \text{ kJ} = 169 \text{ kJ}$$

13.34 First, calculate the number of moles in 33.3 g of water, molar mass 18.02 g/mol.

$$33.0 \text{ g} \times \frac{1 \text{ mol}}{18.02 \text{ g}} = 1.831 \text{ mol}$$

Next, calculate the heat produced.

$$1.831 \text{ mol} \times \frac{40.7 \text{ kJ}}{1 \text{ mol}} = 74.52 \text{ kJ} = 74.5 \text{ kJ}$$

13.35 First, determine the moles of gold (molar mass 196.97 g/mol).

$$26.4 \text{ g} \times \frac{1 \text{ mol}}{196.97 \text{ g}} = 0.1340 \text{ mol}$$

Next, calculate the heat released.

$$0.1340 \text{ mol} \times \frac{310. \text{ kJ}}{1 \text{ mol}} = 41.54 \text{ kJ} = 41.5 \text{ kJ}$$

13.36 First, determine the moles of gold (molar mass 196.97 g/mol).

$$1.0 \text{ g} \times \frac{1 \text{ mol}}{196.97 \text{ g}} = 0.00508 \text{ mol}$$

Next, calculate the heat required.

$$0.00508 \text{ mol} \times \frac{310 \text{ kJ}}{1 \text{ mol}} = 1.57 \text{ kJ} = 1.6 \text{ kJ}$$

13.37 This is a three-step process. The first step is to calculate the heat required to change 12.2 g of ice to liquid water at 0°C. The moles of water are (molar mass 18.02 g)

$$12.2 \text{ g} \times \frac{1 \text{ mol}}{18.02 \text{ g}} = 0.6770 \text{ mol}$$

The heat required for this step is

$$0.6770 \text{ mol} \times \frac{6.01 \text{ kJ}}{1 \text{ mol}} = 4.069 \text{ kJ}$$

The second step is to calculate the heat required to change the water from 0°C to 100°C, a temperature change of 100°C.

$$12.2 \text{ g} \times \frac{4.184 \text{ J}}{1 \text{ g} \cdot °C} \times 100 °C = 5104 \text{ J} = 5.104 \text{ kJ}$$

The third step is to calculate the heat required to change the water at 100°C to steam.

$$0.6770 \text{ mol} \times \frac{40.7 \text{ kJ}}{1 \text{ mol}} = 27.55 \text{ kJ}$$

The total heat required for the process is

$$4.069 \text{ kJ} + 5.104 \text{ kJ} + 27.55 \text{ kJ} = 36.\underline{7}2 \text{ kJ} = 36.7 \text{ kJ}$$

13.38 This is a three-step process. The first step is to calculate the heat liberated when 53 g of steam at 100°C is converted to liquid water. The moles of water are (molar mass 18.02 g)

$$53 \text{ g} \times \frac{1 \text{ mol}}{18.02 \text{ g}} = 2.94 \text{ mol}$$

The heat liberated from this step is

$$2.94 \text{ mol} \times \frac{40.7 \text{ kJ}}{1 \text{ mol}} = 120 \text{ kJ}$$

The second step is to calculate the heat liberated when liquid water is cooled from 100°C to 0°C, a temperature change of 100°C.

$$53 \text{ g} \times \frac{4.184 \text{ J}}{1 \text{ g} \cdot °C} \times 100 °C = 22,200 \text{ J} = 22.2 \text{ kJ}$$

LIQUIDS, SOLIDS, AND ATTRACTIONS BETWEEN MOLECULES ■ 185

The third step is to calculate the heat liberated when the water is converted into ice at 0°.

$$2.94 \text{ mol} \times \frac{6.01 \text{ kJ}}{1 \text{ mol}} = 17.7 \text{ kJ}$$

Thus, the total heat liberated in the process is

$$120 \text{ kJ} + 22.2 \text{ kJ} + 17.7 \text{ kJ} = 1\underline{6}0. \text{ kJ} = 1.6 \times 10^2 \text{ kJ}$$

13.39 This is a two-step process. First, find the heat needed to change 33.3 g of water at 37°C to water at 100°C, a temperature change of 63°C.

$$33.3 \text{ g} \times \frac{4.184 \text{ J}}{1 \text{ g} \cdot °C} \times 63 \text{ °C} = 8,780 \text{ J} = 8.78 \text{ kJ}$$

The second step is to calculate the heat needed to vaporize the water at 100°C. The moles of water are (molar mass 18.02 g)

$$33.3 \text{ g} \times \frac{1 \text{ mol}}{18.02 \text{ g}} = 1.848 \text{ mol}$$

The heat needed is

$$1.848 \text{ mol} \times \frac{40.7 \text{ kJ}}{1 \text{ mol}} = 75.21 \text{ kJ}$$

The total heat required for the process is

$$8.78 \text{ kJ} + 75.21 \text{ kJ} = 83.\underline{9}9 \text{ kJ} = 84.0 \text{ kJ}$$

13.40 This is a two-step process. First, find the heat liberated when 45 g of steam is converted into liquid water at 100°C. The moles of water are (molar mass 18.02 g)

$$45 \text{ g} \times \frac{1 \text{ mol}}{18.02 \text{ g}} = 2.50 \text{ mol}$$

The heat liberated is

$$2.50 \text{ mol} \times \frac{40.7 \text{ kJ}}{1 \text{ mol}} = 102 \text{ kJ}$$

186 ■ CHAPTER 13

The second step is to calculate the heat liberated when the liquid water cools from 100°C to 43°C, a temperature change of 57°C.

$$45 \text{ g} \times \frac{4.184 \text{ J}}{1 \text{ g} \cdot °C} \times 57 °C = 10,700 \text{ J} = 10.7 \text{ kJ}$$

The total heat liberated in the process is

$$102 \text{ kJ} + 10.7 \text{ kJ} = 113 \text{ kJ} = 1.1 \times 10^2 \text{ kJ}$$

13.41 a. The only attractive forces present in radon (Rn) are dispersion forces.

b. Hydrogen fluoride (HF) is polar and has dipole-dipole forces as well as dispersion forces. In addition, since the hydrogen is bonded to a fluorine atom, there will be hydrogen bonding.

c. Hydrogen iodide (HI) is polar and has dipole-dipole forces as well as dispersion forces.

13.42 a. Hydrogen sulfide (H_2S) is polar and has dipole-dipole forces as well as dispersion forces.

b. Water (H_2O) is polar and has dipole-dipole forces as well as dispersion forces. In addition, since the hydrogen is bonded to oxygen, there will be hydrogen bonding.

c. Iodine (I_2) is nonpolar so the only attractive forces are dispersion forces.

13.43 Dispersion forces increase with increasing molecular weight. Since each of the molecules is a nonpolar tetrahedral molecule with only dispersion forces present, the order is CCl_4 < $SiCl_4$ < $GeCl_4$.

13.44 The noble gases have only dispersion forces present. Therefore, the forces increase as the atomic weight increases. Thus, the order is He < Ar < Rn.

13.45 Both CH_4 and CCl_4 are nonpolar tetrahedral molecules so dispersion forces are the only attractive forces present. Since dispersion forces increase with increasing molar mass, the lower vapor pressure would be for CCl_4, where the dispersion forces are stronger.

13.46 The interhalogen compound with the lower vapor pressure would be the one with the stronger intermolecular forces. Since both compounds are polar, the larger dipole-dipole forces would exist for the compound with the higher electronegativity difference, BrF, and it would have the lower vapor pressure.

13.47 Boiling point increases with increasing intermolecular forces. This is because the stronger the attractive forces the lower the vapor pressure and therefore the higher the temperature necessary to reach boiling. Since the three compounds are all nonpolar tetrahedral molecules and only have dispersion forces present, the dispersion forces increase with increasing molar mass. Thus, the order of increasing boiling point is $CCl_4 < SiCl_4 < GeCl_4$.

13.48 The noble gases have only dispersion forces. Since dispersion forces increase with increasing molar mass, the boiling point also increases with increasing molar mass. Thus, the order is He < Ar < Rn.

13.49 Hydrogen selenide (H_2Se) and water (H_2O) are both polar and have dipole-dipole forces. However, water also has hydrogen bonding and hydrogen selenide does not. Therefore, water has the stronger forces and the higher boiling point.

13.50 Phosphine (PH_3) and ammonia (NH_3) are both polar and have dipole-dipole forces. However, ammonia also has hydrogen bonding and phosphine does not. Therefore, ammonia has the stronger forces and the higher boiling point.

13.51 P_4 is a molecular substance. Therefore, it freezes as a <u>molecular solid</u>.

13.52 O_2 molecules will form a <u>molecular solid</u>.

13.53 Barium chloride ($BaCl_2$) is an <u>ionic solid</u>.

13.54 Titanium is a metallic element. Therefore, it exists as a <u>metallic solid</u>.

13.55 Since both water (H_2O) and hydrogen selenide (H_2Se) are molecular substances, they would both freeze as molecular solids. The stronger attractive forces are in water due to hydrogen bonding. Thus, water has the higher boiling point and the higher melting point.

13.56 Both H_2Se and H_2Te are polar molecules and have dipole-dipole forces. Neither compound will have hydrogen bonding. The melting points increase with increasing molar mass. Thus H_2Te will have the higher melting point.

13.57 Nitrogen (N_2) is a gas at room temperature while phosphorus (P_4) is a solid. Therefore, phosphorus has the higher melting point.

13.58 Ozone (O_3) is polar while oxygen (O_2) is not. Therefore, since both are molecular substances, ozone has the higher boiling point and also the higher melting point.

Solutions to Additional Problems

Note on significant figures: The first time the final answer is written, it is written first with one nonsignificant figure. The least significant digit is also underlined. The final answer is then rounded to the correct number of significant figures but the least significant digit is no longer underlined.

13.59 We must first calculate the heat that is released when 525 g of water freezes at 0°C. The number of moles of water is

$$525 \text{ g} \times \frac{1 \text{ mol}}{18.02 \text{ g}} = 29.13 \text{ mol}$$

The heat released is

$$29.13 \text{ mol} \times \frac{6.01 \text{ kJ}}{1 \text{ mol}} = 175.1 \text{ kJ}$$

Since the molar heat of fusion for CCl_2F_2 is 17.4 kJ/mol, the moles of this compound required to absorb the heat is

$$175.1 \text{ kJ} \times \frac{1 \text{ mol}}{17.4 \text{ kJ}} = 10.06 \text{ mol}$$

The molar mass of CCl_2F_2 is 120.91 g/mol. Therefore, the mass required is

$$10.06 \text{ mol} \times \frac{120.91 \text{ g}}{1 \text{ mol}} = 12\underline{1}6 \text{ g} = 1.22 \times 10^3 \text{ g}$$

13.60 The molar mass of ammonia is 17.03 g/mol. The number of moles in 1.00 kg of ammonia is

$$1.00 \text{ kg} \times \frac{1000 \text{ g}}{1 \text{ kg}} \times \frac{1 \text{ mol}}{17.03 \text{ g}} = 58.72 \text{ mol}$$

The amount of heat that the ammonia can absorb is

$$58.72 \text{ mol} \times \frac{23.4 \text{ kJ}}{1 \text{ mol}} = 1374 \text{ kJ}$$

Since the molar heat of fusion of water is 6.01 kJ/mol, the moles of water that can be frozen is

$$1374 \text{ kJ} \times \frac{1 \text{ mol}}{6.01 \text{ kJ}} = 228.6 \text{ mol}$$

The molar mass of water is 18.01 g/mol. Therefore, the mass of water is

$$228.6 \text{ mol} \times \frac{18.01 \text{ g}}{1 \text{ mol}} = 4118 \text{ g} = 4.12 \times 10^3 \text{ g}$$

13.61 The heat of vaporization is a measure of the energy required to overcome the attractive forces in the substance. The stronger the forces, the higher the heat of vaporization. Since both Cl_2 and H_2 are nonpolar and only have dispersion forces present, the forces increase with increasing molecular mass. Thus Cl_2 should have the higher heat of vaporization, and it does.

13.62 Since CH_3OH is polar and Ne is nonpolar, the stronger intermolecular forces are in CH_3OH. There is also hydrogen bonding present. Therefore, CH_3OH should have the higher heat of vaporization, and it does.

14. SOLUTIONS

■ Solutions to Exercises

<u>Note on significant figures:</u> The first time the final answer is written, it is written first with one nonsignificant figure. The least significant digit is also underlined. The final answer is then rounded to the correct number of significant figures but the least significant digit is no longer underlined.

14.1 a. Na_2SO_4: Sodium salts are generally soluble; no exceptions are listed in Table 14.2 so sodium sulfate is soluble.

  b. AgCl: Chlorides are usually soluble, but silver chloride is an exception. Silver chloride is insoluble.

14.2 Substitute into the defining equation for mass percent of solute:

$$\text{Mass percent of solute} = \frac{\text{mass of solute}}{\text{mass of solution}} \times 100\%$$

$$= \frac{5.4 \text{ g}}{83.5 \text{ g}} \times 100\% = 6.\underline{4}7\% = 6.5\%$$

14.3 You first calculate the mass of the solution:

Mass of solution = mass of solute + mass of solvent = 5.4 g + 83.5 g = 88.9 g

Next, you calculate the mass percent of the solute:

$$\frac{5.4 \text{ g}}{88.9 \text{ g}} \times 100\% = 6.\underline{0}7\% = 6.1\%$$

SOLUTIONS ■ 191

14.4 Start with the equation for mass of solute:

$$\text{Mass of solute} = \frac{\text{mass percent of solute} \times \text{mass of solution}}{100\%}$$

Substitute into this equation to obtain the mass of solute:

$$\text{Mass of solute} = \frac{5.0\% \times 42.5 \text{ g}}{100\%} = 2.13 \text{ g} = 2.1 \text{ g}$$

Mass of water = 42.5 g - 2.13 g = 40.38 g = 40.4 g

14.5 First, convert the volume to liters: 38.0 mL is 0.0380 L. Next, use the definition of molarity:

$$\text{Molarity} = \frac{\text{moles of solute}}{\text{liters of solution}} = \frac{0.0350 \text{ mol}}{0.0380 \text{ L}} = 0.9211 \text{ M} = 0.921 \text{ M}$$

14.6 In order to obtain the moles of $NiSO_4$, you need the molecular weight, 154.75 amu. Thus, 1 mol of $NiSO_4$ = 154.75 g. Hence,

$$1.58 \text{ g } NiSO_4 \times \frac{1 \text{ mol } NiSO_4}{154.75 \text{ g } NiSO_4} = 0.01021 \text{ mol } NiSO_4$$

The volume of the solution is 61.2 mL, or 0.0612 L. Thus,

$$\text{Molarity} = \frac{0.01021 \text{ mol}}{0.0612 \text{ L}} = 0.1668 \text{ M} = 0.167 \text{ M}$$

14.7 First, convert the volume to moles using the molarity (250 mL = 0.250 L).

$$0.250 \text{ L } AgNO_3 \text{ solution} \times \frac{0.150 \text{ mol } AgNO_3}{1 \text{ L } AgNO_3 \text{ solution}} = 0.03750 \text{ mol } AgNO_3$$

Next, use the formula weight of $AgNO_3$ (169.88 amu) to find the mass.

$$0.03750 \text{ mol } AgNO_3 \times \frac{169.88 \text{ g } AgNO_3}{1 \text{ mol } AgNO_3} = 6.371 \text{ g} = 6.37 \text{ g}$$

14.8 First, obtain the moles of NaOH. The volume is 48.1 mL or 0.0481 L.

$$0.0481 \text{ L NaOH solution} \times \frac{0.148 \text{ mol NaOH}}{1 \text{ L NaOH solution}} = 0.007119 \text{ mol NaOH}$$

The molecular weight of $HC_2H_3O_2$ is 60.03 amu. Using the balanced chemical equation gives:

$$0.007119 \text{ mol NaOH} \times \frac{1 \text{ mol } HC_2H_3O_2}{1 \text{ mol NaOH}} \times \frac{60.03 \text{ g } HC_2H_3O_2}{1 \text{ mol } HC_2H_3O_2}$$

$$= 0.4274 \text{ g} = 0.427 \text{ g}$$

14.9 The molarity of the solution is

$$\text{molarity} = \frac{0.244 \text{ mol HCl}}{1 \text{ L solution}} \times 0.0282 \text{ L solution} \times \frac{1 \text{ mol Ba(OH)}_2}{2 \text{ mol HCl}} \times$$

$$\frac{1}{0.0250 \text{ L solution}} = 0.1376 \text{ M} = 0.138 \text{ M}$$

14.10 The amount of $AgNO_3$ in the dilute solution is

$$1.00 \text{ L AgNO}_3 \text{ solution} \times \frac{0.10 \text{ mol AgNO}_3}{1 \text{ L AgNO}_3 \text{ solution}} = 0.100 \text{ mol AgNO}_3$$

This is also the moles of $AgNO_3$ in the concentrated solution. Convert this to liters of solution using the molarity of the solution.

$$0.100 \text{ mol AgNO}_3 \times \frac{1 \text{ L solution}}{2.5 \text{ mol AgNO}_3} = 0.0400 \text{ L} = 0.040 \text{ L} = 40. \text{ mL}$$

14.11 Substitute into the dilution equation to get

$$1.2 \text{ mol/L} \times V_i = 0.25 \text{ mol/L} \times 0.50 \text{ L}$$

Solving for V_i gives

$$V_i = \frac{0.25 \text{ mol/L} \times 0.50 \text{ L}}{1.2 \text{ mol/L}} = 0.1\underline{0}4 \text{ L} = 0.10 \text{ L}$$

Therefore, 0.10 L, or 1.0×10^2 mL, of concentrated solution should be used to prepare the dilute solution.

14.12 First, find the moles of ethylene glycol, $C_2H_6O_2$, molecular weight 62.07 amu.

$$34.2 \text{ g } C_2H_6O_2 \times \frac{1 \text{ mol } C_2H_6O_2}{62.07 \text{ g } C_2H_6O_2} = 0.5510 \text{ mol } C_2H_6O_2$$

Using the molality formula with 250.0 g, or 0.2500 kg, of solvent gives

$$\text{molality} = \frac{0.5510 \text{ mol } C_2H_6O_2}{0.2500 \text{ kg solvent}} = 2.2\underline{0}4 \text{ m} = 2.20 \text{ m}$$

14.13 You must first find the molality of the solution. Sucrose, $C_{12}H_{22}O_{11}$, has a molecular weight 342.30 amu. Find the moles of sucrose

$$23.4 \text{ g } C_{12}H_{22}O_{11} \times \frac{1 \text{ mol } C_{12}H_{22}O_{11}}{342.30 \text{ g } C_{12}H_{22}O_{11}} = 0.06836 \text{ mol } C_{12}H_{22}O_{11}$$

Using the molality formula with 125 g, or 0.125 kg, solvent gives

$$\text{molality} = \frac{0.06836 \text{ mol } C_{12}H_{22}O_{11}}{0.125 \text{ kg solvent}} = 0.5469 \text{ m}$$

You can now find the temperature change, ΔT_f.

$$\Delta T_f = K_f m = (1.86°C/m) \times (0.5469 \text{ m}) = 1.017°C$$

Therefore, the freezing point of the solution is $0.00°C - 1.017°C = -1.0\underline{1}7°C = -1.02°C$

14.14 The boiling point of the solution in Exercise 13 (molality = 0.5469 m) is found by first finding ΔT_b:

$$\Delta T_b = K_b m = (0.52°C/m) \times (0.5469\ m) = 0.284°C$$

The boiling point is 100.00°C + 0.284°C = 100.2$\underline{8}$4°C = 100.28°C

14.15 $\Delta T_f = 0.000°C - (-0.242°C) = 0.242°C$

The molality of the solution can now be found

$$\text{molality} = \frac{\Delta T_f}{K_f} = \frac{0.242°C}{1.86°C/m} = 0.1301\ m$$

You can now convert the kilograms of solvent (135 g = 0.135 kg water) into moles of solute:

$$0.135\ \text{kg solvent} \times \frac{0.1301\ \text{mol solute}}{1\ \text{kg solvent}} = 0.01756\ \text{mol solute}$$

Therefore, 6.00 g of solute equals 0.01756 mol solute, so

$$\text{molar mass unknown} = \frac{6.00\ g}{0.01756\ \text{mol}} = 34\underline{1}.7\ g/mol = 342\ g/mol,\ \text{or 342 amu}$$

■ Answers to Questions to Test Your Reading

14.1 A <u>solution</u> is a homogeneous mixture of two or more substances. When one of these substances is a liquid and the others are solids or gases, the liquid is the <u>solvent</u> and the solids and gases are the <u>solutes</u>. In a sodium chloride and water mixture, the water is the solvent, sodium chloride is the solute, and the mixture is the solution. Alternatively, the substance present in the greater amount is the solvent.

14.2 Two liquids that mix completely in one another to form a solution, no matter what the proportions of liquids, are said to be <u>miscible</u>.

14.3 Two liquids that separate into two distinct layers after mixing are said to be <u>immiscible</u>. No matter how much you stir, they will not mix together.

SOLUTIONS ■ 195

14.4 a. Air is an example of a gaseous solution. It contains 78% N_2, 21% O_2, and also Ar, CO_2, and H_2O in trace amounts.

b. An example of a liquid solvent and gaseous solute is carbonation in beverages. This is carbon dioxide gas dissolved in a water solution.

c. A mixture of the appropriate amounts of solid sodium with solid potassium forms a liquid solution.

d. An alloy, such as jewelry gold, which is a mixture of gold and silver, is a solid solution.

e. Gasoline is an example of a nonaqueous solution.

14.5 a. A mixture of helium and oxygen is a gaseous solution. The solute and solvent are both gases.

b. Brass (copper and zinc) is a solid solution, an alloy. The solute and the solvent are both solids.

c. Antifreeze and water is a liquid solution. The solute and the solvent are both liquids.

d. Brine is a mixture of salt in water, an aqueous solution. The solute is a solid and the solvent is a liquid, water.

e. Chlorine (in the gaseous state) dissolves in water as a gaseous solute and a liquid solvent.

14.6 A compound is a pure substance of fixed composition by mass. An example is water, 88.8% oxygen and 11.2% hydrogen by mass. A solution is two or more substances mixed in variable proportions. Also, a solution can be separated into its pure components by a physical process like distilling, while a compound can only be separated into elements by a chemical process.

14.7 In a bucket of sand and water, the sand will settle out because the particles are large enough to be affected by gravitational forces. Since sand is more dense than water it falls to the bottom of the bucket. Solute molecules dissolve to form individual molecules or ions that are stabilized by the solvent molecules so that gravity no longer affects them as a principal force. The individual particles are in constant, random motion due to collisions with other particles.

14.8 The solubility of lithium hydroxide in water at 20°C is 12.4 g per 100 mL of water. This means that a 100 mL volume of water dissolves a maximum of 12.4 g of lithium hydroxide at 20°C. Any more lithium hydroxide that you add would settle to the bottom.

14.9 The solution you obtain after you have dissolved the maximum amount of solute in a solvent at a given temperature is said to be a <u>saturated solution</u>. If you take a small quantity of the solute and add it to the saturated solution, it will simply fall to the bottom of the vessel. In an unsaturated solution, it would dissolve.

14.10 The solubility of lithium hydroxide in water at 20°C is 12.4 g. Since 11.0 g is less than the maximum amount per 100 mL of water, it is an unsaturated solution.

14.11 A <u>supersaturated solution</u> is a solution that contains more of a solute than is contained in a saturated solution. If you add a crystal to a supersaturated solution, it initiates a crystallization process resulting in a saturated solution, with excess solute settling to the bottom as crystals. In a saturated solution, if you add a crystal, it will simply fall to the bottom, without dissolving.

14.12 Since the solubility of potassium dichromate ($K_2Cr_2O_7$) in 100 mL of water is 11.7 g, this is the maximum amount of the material that will dissolve. The excess potassium dichromate that would remain undissolved would be 15.0 g - 11.7 g = 3.3 g.

14.13 In a saturated solution of potassium dichromate, there is an equilibrium that exists where the process of dissolution (in which ions leave the potassium dichromate crystals and enter the solution) and crystallization (in which ions in the solution attach themselves to a crystal). The equilibrium can be represented by

$$K_2Cr_2O_7(s) \rightleftharpoons 2\ K^+(aq) + Cr_2O_7^{2-}(aq)$$

Even though the amount of potassium dichromate in the solution remains constant, the process of dissolution has not stopped. The rate at which the ions in the solution come back and reattach themselves to crystals becomes equal to the rate at which ions are leaving the crystal. Ions are constantly leaving and returning to the crystals, but at the same rate.

14.14 One reason that some compounds dissolve better than others is due to the attractive forces between the water molecules and the ions. This attractive force is in competition with the force of attraction between ions in the solid. Whichever force is stronger dominates. If the force between ions and water is stronger, the compound dissolves more readily than if the ionic forces are stronger.

SOLUTIONS ■ 197

14.15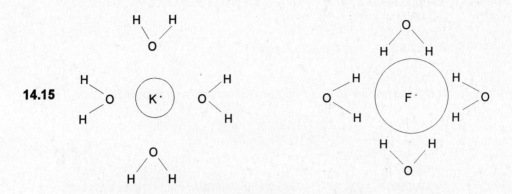

The difference between the two is that for positive ions like K^+, the negative end (the oxygen end) of the polar water molecule is closest to the ion whereas for negative ions like F^-, it is the positive end (the hydrogen end) that is the closest.

14.16 Carbon tetrachloride (CCl_4) is nonpolar. This means that it has no positive or negative end so it will not dissolve in polar liquids. Therefore, it should be more soluble in gasoline, another nonpolar liquid, than in water. The general rule is like dissolves like. Polar substances dissolve in one another and nonpolar substances dissolve in one another, but polar and nonpolar do not mix.

14.17 If you warm up a carbonated beverage, it will go flat. This is because carbon dioxide (carbonation) is more soluble at lower temperatures than at higher temperatures. Thus, as you warm up the beverage, carbon dioxide will bubble out.

14.18 If a saturated solution of sodium nitrate is heated up, it will not remain saturated. Sodium nitrate is more soluble at higher temperatures, so as you heat up the solution, you must dissolve more of the sodium nitrate in order to achieve a saturated solution again. In general, solubility increases with temperature.

14.19 The solubility of a gas, like nitrogen, increases with increasing pressure. Since the pressure exerted on a diver is higher at the deeper depths, more nitrogen dissolves in the blood at the deeper depths. As the diver ascends, the pressure decreases, and the extra nitrogen begins to bubble out. If it occurs too rapidly, it causes the bends.

14.20a. A 4.0% by mass aqueous solution means that for every 100 g of the solution there is 4.0 g of the solute and 96.0 g of water.

   b. A concentrated, 14.5 M ammonia solution has 14.5 moles of ammonia for every liter of solution.

   c. An aqueous sucrose solution that has a concentration of 0.15 m means that there is 0.15 moles of sucrose dissolved in one kilogram of water.

CHAPTER 14

14.21 A conversion factor that converts from liters of solution to moles H_2SO_4 would be

$$\frac{0.18 \text{ mol } H_2SO_4}{1 \text{ liter } H_2SO_4 \text{ solution}}$$

14.22 A conversion factor that converts from kilograms solvent to moles of H_2SO_4 would be

$$\frac{0.18 \text{ mol } H_2SO_4}{1 \text{ kg water}}$$

14.23 Molarity is convenient when you dispense molar amounts of substances by measuring out volumes of solutions, and measuring volumes is simple and very accurate.

14.24 As you increase the temperature of a solution, its volume increases. However, the moles of solute remains constant. This causes a decrease in the molarity (mol/L) of the solution as the temperature increases. On the other hand, the mass of the solvent is not a function of the temperature so the molality (mol/kg) remains constant.

14.25 A titration would be carried out as follows. A measured volume of citric acid solution would be placed in a flask and an indicator is added to indicate the completion of the reaction. A sodium hydroxide (NaOH) solution of known molarity would then be slowly added to the acid solution. When the indicator changes color, the reaction is complete and the volume of the sodium hydroxide solution that was added can now be measured.

The calculation of the mass of citric acid could be determined as follows. First, use the molarity formula to convert liters of NaOH solution to moles of NaOH. Next, use the balanced chemical equation to convert moles of NaOH to moles of citric acid. Finally, multiply the moles of citric acid by the molar mass of citric acid to find the mass of citric acid in the solution.

14.26 To prepare a solution of a specific desired molarity from a more concentrated solution of known molarity, you can use either of the following two methods to determine the volume of the concentrated solution that is needed. The first method is the conversion factor method and the second method is using the dilution equation, $M_iV_i = M_fV_f$. In either case, the volume of concentrated solution is determined. Sufficient water is then added to obtain the final volume of solution.

14.27 A 1.0 L solution of 2.0 M HCl contains 2.0 moles of HCl. Since a 1.0 M solution of HCl can only contain 1.0 mole of HCl, it would take two liters of solution to contain the 2.0 moles of HCl. Therefore, the final volume of the solution would be 2.0 L.

SOLUTIONS ■ 199

14.28 The freezing point depression, ΔT_f, is the colligative property of a solution equal to the freezing point of the pure solvent minus the freezing point of the solution.

$$\Delta T_f = \text{f.p. pure solvent} - \text{f.p. solution}$$

It is also equal to the freezing point depression constant for the solvent, K_f, multiplied by the molality (for a nonvolatile, nonionic solid). For water, $K_f = 1.86°C/m$

$$\Delta T_f = K_f m$$

Therefore, when m = 1.0, ΔT_f = 1.9°C, and when m = 2.0, ΔT_f = 2.0 x 1.86°C, or 3.7°C.

14.29 For an aqueous solution, $K_f = 1.86°C/m$. For a solution of either glucose or sucrose with concentration of 0.10 m, the freezing point depression is

$$\Delta T_f = K_f m = (1.86°C/m) \times 0.10\ m = 0.19°C$$

Thus, the freezing point of the solution would be -0.19°C (pure water freezes at 0.00°C) K_b for water is 0.52°C/m. The boiling point elevation of a solution of either glucose or sucrose with the same concentration is

$$\Delta T_b = K_b m = (0.52°C/m) \times 0.10\ m = 0.052°C$$

Thus, the boiling point of the solution would be 100.052°C (the boiling point of pure water is 100.000°C).

14.30 Since NaCl forms ions in solution and urea doesn't, there are more solute particles in a NaCl solution than in a solution of urea of the same molality. Since the freezing point depression is a function of the number of solute particles in the solution, the effect is greater for NaCl.

14.31 By the process of osmosis, water would flow out of the greens into the vinegar solution until the concentration of the solute particles is diluted. The effect is that the greens will lose their moisture and wilt.

14.32 A solution that has an osmotic pressure equal to that of blood plasma is <u>isotonic</u> with blood plasma. An example is a 0.85% by mass solution of sodium chloride. This is called a physiological saline solution.
If red blood cells are placed in a more concentrated solution, osmosis would occur and water would pass out of the blood cells and they would shrink.

Solutions to Practice Problems

Note on significant figures: The first time the final answer is written, it is written first with one nonsignificant figure. The least significant digit is also underlined. The final answer is then rounded to the correct number of significant figures but the least significant digit is no longer underlined.

14.33a. All nitrates are soluble without any exceptions, so $Pb(NO_3)_2$ is soluble.

b. All potassium salts are soluble so K_3PO_4 is soluble.

c. Most chlorides are soluble but lead(II) chloride is an exception. $PbCl_2$ is insoluble.

d. Most phosphates are insoluble, so $Ca_3(PO_4)_2$ is insoluble.

e. All potassium salts are soluble, so KOH is soluble.

14.34a. Most sulfates are soluble but strontium sulfate is an exception. $SrSO_4$ is insoluble.

b. Most chlorides are soluble. $CaCl_2$ is soluble.

c. Most hydroxides are insoluble. $Ni(OH)_2$ is insoluble.

d. All ammonium salts are soluble. NH_4Cl is soluble.

e. All nitrates are soluble. $LiNO_3$ is soluble.

14.35a. $\dfrac{15 \text{ g } Pb(NO_3)_2}{45 \text{ g solution}} \times 100\% = 3\underline{3}.3\% = 33\%$

b. $\dfrac{13.6 \text{ g } Na_2SO_4}{125 \text{ g solution}} \times 100\% = 10.\underline{8}8\% = 10.9\%$

14.36a. $\dfrac{0.236 \text{ g } NH_4NO_3}{24.1 \text{ g solution}} \times 100\% = 0.97\underline{9}3\% = 0.979\%$

b. $\dfrac{7.1 \text{ g } HC_2H_3O_2}{134 \text{ g solution}} \times 100\% = 5.\underline{3}0\% = 5.3\%$

SOLUTIONS ■ 201

14.37 First, calculate the mass of the solution:

mass solution = 2.34 g + 255 g = 257.34 g

Then, calculate the mass percent:

$$\frac{2.34 \text{ g Pb(NO}_3)_2}{257.34 \text{ g solution}} \times 100\% = 0.909\underline{3}\% = 0.909\%$$

14.38 First, calculate the mass of the solution:

mass solution = 5.0 g + 85 g = 90.0 g

Then, calculate the mass percent:

$$\frac{5.0 \text{ g NaCl}}{90.0 \text{ g solution}} \times 100\% = 5.\underline{5}5\% = 5.6\%$$

14.39 Use the equation:

$$\text{Mass of solute} = \frac{\text{mass percent of solute} \times \text{mass of solution}}{100\%}$$

Substitute in to get the mass of $CuSO_4$:

$$\frac{5.0\% \times 425 \text{ g}}{100\%} = 2\underline{1}.3 \text{ g} = 21 \text{ g}$$

The mass of water is

mass water = 425 g - 21.3 g = 40\underline{3}.7 g = 404 g

14.40 The mass of NH_4Cl is

$$\frac{9.5\% \times 124 \text{ g}}{100\%} = 1\underline{1}.8 \text{ g} = 12 \text{ g}$$

The mass of water is

mass water = 124 g - 11.8 g = 11\underline{2}.2 g = 112 g

202 ■ CHAPTER 14

14.41 Rearrange the mass percent equation to get:

$$\text{Mass of solution} = \frac{\text{Mass of solute} \times 100\%}{\text{Mass percent of solute}}$$

Substitute in to get the mass of the solution:

$$\frac{5.0 \text{ g} \times 100\%}{7.35\%} = 68.0 \text{ g} = 68 \text{ g}$$

14.42 (See problem 41) The mass of the solution is

$$\frac{0.257 \text{ g} \times 100\%}{5.00\%} = 5.140 \text{ g} = 5.14 \text{ g}$$

14.43 First, find the mass of NaOH (molecular weight 40.00 amu). So, one mole of NaOH weighs 40.00 g.

$$0.145 \text{ mol NaOH} \times \frac{40.00 \text{ g NaOH}}{1 \text{ mol NaOH}} = 5.800 \text{ g NaOH}$$

Next, find the mass of the solution:

mass solution = 5.800 g + 200. g = 205.800 g

Now, find the mass percent of NaOH:

$$\frac{5.800 \text{ g NaOH}}{205.800 \text{ g solution}} \times 100\% = 2.818\% = 2.82\%$$

14.44 First, find the mass of $CuSO_4$ (molecular weight 159.61 amu). So, one mole of $CuSO_4$ weighs 159.61 g.

$$0.100 \text{ mol CuSO}_4 \times \frac{159.61 \text{ g CuSO}_4}{1 \text{ mol CuSO}_4} = 15.961 \text{ g CuSO}_4$$

Next, find the mass of the solution (1.00 kg = 1000 g).

mass solution = 15.961 g + 1000 g = 1015.961 g

Now, find the mass percent of $CuSO_4$:

$$\frac{15.961 \text{ g CuSO}_4}{1015.961 \text{ g solution}} \times 100\% = 1.571\% = 1.57\%$$

14.45 a. 250. mL is 0.250 L. So, the molarity is:

$$\frac{0.0450 \text{ mol Cu(NO}_3)_2}{0.250 \text{ L}} = 0.1800 \text{ M} = 0.180 \text{ M}$$

b. 455 mL is 0.455 L. So, the molarity is:

$$\frac{0.250 \text{ mol NH}_4\text{NO}_3}{0.455 \text{ L}} = 0.5494 \text{ M} = 0.549 \text{ M}$$

14.46 50.0 mL is 0.0500 L.

a. $\dfrac{0.0235 \text{ mol NaNO}_3}{0.0500 \text{ L}} = 0.4700 \text{ M} = 0.470 \text{ M}$

b. $\dfrac{0.178 \text{ mol CuCl}_2}{0.0500 \text{ L}} = 3.560 \text{ M} = 3.56 \text{ M}$

14.47 a. First, find the moles of $Ba(NO_3)_2$ (molecular weight 261.32 amu).

$$0.418 \text{ g Ba(NO}_3)_2 \times \frac{1 \text{ mol Ba(NO}_3)_2}{261.32 \text{ g Ba(NO}_3)_2} = 0.001600 \text{ mol Ba(NO}_3)_2$$

Since 25.0 mL is 0.0250 L, the molarity is:

$$\frac{0.001600 \text{ mol Ba(NO}_3)_2}{0.0250 \text{ L}} = 0.06398 \text{ M} = 0.0640 \text{ M}$$

b. First, find the moles of $NaHSO_4$ (molecular weight 120.06 amu).

$$23.4 \text{ mg} \times \frac{1 \text{ g}}{1000 \text{ mg}} \times \frac{1 \text{ mol}}{120.06 \text{ g}} = 1.949 \times 10^{-4} \text{ mol}$$

Since 50.0 mL is 0.0500 L, the molarity is:

$$\frac{1.949 \times 10^{-4} \text{ mol}}{0.0500 \text{ L}} = 0.003898 \text{ M} = 0.00390 \text{ M}$$

14.48a. First, find the moles of $CaCl_2$ (molecular weight 110.98 amu).

$$28.5 \text{ g} \times \frac{1 \text{ mol}}{110.98 \text{ g}} = 0.2568 \text{ mol}$$

Since 755 mL is 0.755 L, the molarity is:

$$\frac{0.2568 \text{ mol}}{0.755 \text{ L}} = 0.3401 \text{ M} = 0.340 \text{ M}$$

b. First, find the moles of $AgNO_3$ (molecular weight 169.91).

$$85.6 \text{ mg} \times \frac{1 \text{ g}}{1000 \text{ mg}} \times \frac{1 \text{ mol}}{169.91 \text{ g}} = 5.038 \times 10^{-4} \text{ mol}$$

Since 25.0 mL is 0.0250 L, the molarity is:

$$\frac{5.038 \times 10^{-4} \text{ mol}}{0.0250 \text{ L}} = 0.02015 \text{ M} = 0.0202 \text{ M}$$

14.49 The molecular weight of K_2CrO_4 is 194.20 amu. Since 385 mL is 0.385 L, the mass of K_2CrO_4 needed is

$$0.385 \text{ L solution} \times \frac{0.0750 \text{ mol } K_2CrO_4}{1 \text{ L solution}} \times \frac{194.20 \text{ g}}{1 \text{ mol } K_2CrO_4} = 5.607 \text{ g} = 5.61 \text{ g}$$

14.50 The molecular weight of $KMnO_4$ is 158.04 amu. Since 25.0 mL is 0.0250 L, the mass of $KMnO_4$ needed is

$$0.0250 \text{ L solution} \times \frac{0.300 \text{ mol } KMnO_4}{1 \text{ L solution}} \times \frac{158.04 \text{ g}}{1 \text{ mol } KMnO_4} = 1.185 \text{ g} = 1.19 \text{ g}$$

14.51 $\quad 0.150 \text{ mol } Na_2CO_3 \times \dfrac{1 \text{ L solution}}{0.265 \text{ mol } Na_2CO_3} \times \dfrac{1000 \text{ mL}}{1 \text{ L}} = 566.0 \text{ mL} = 566 \text{ mL}$

SOLUTIONS 205

14.52 $0.0450 \text{ mol } H_2SO_4 \times \dfrac{1 \text{ L solution}}{0.0850 \text{ mol } H_2SO_4} \times \dfrac{1000 \text{ mL}}{1 \text{ L}} = 529.4 \text{ mL} = 529 \text{ mL}$

14.53 First, find the moles of HCl using the molarity. Since 38.9 mL is 0.0389 L, you get:

$$0.0389 \text{ L solution} \times \dfrac{0.115 \text{ mol HCl}}{1 \text{ L solution}} = 0.004474 \text{ mol HCl}$$

The molecular weight of $CaCO_3$ is 100.09 amu. Therefore, the mass of $CaCO_3$ in the sample is

$$0.004474 \text{ mol HCl} \times \dfrac{1 \text{ mol } CaCO_3}{2 \text{ mol HCl}} \times \dfrac{100.09 \text{ g } CaCO_3}{1 \text{ mol } CaCO_3} = 0.2239 \text{ g} = 0.224 \text{ g}$$

14.54 First, find the moles of sulfuric acid using the molarity. Since 25.8 mL is 0.0258 L, you get

$$0.0258 \text{ L solution} \times \dfrac{0.283 \text{ mol } H_2SO_4}{1 \text{ L solution}} = 0.007301 \text{ mol } H_2SO_4$$

The molecular weight of NH_3 is 17.03 amu. Therefore, the mass of ammonia in the sample is

$$0.007301 \text{ mol } H_2SO_4 \times \dfrac{2 \text{ mol } NH_3}{1 \text{ mol } H_2SO_4} \times \dfrac{17.03 \text{ g } NH_3}{1 \text{ mol } NH_3} = 0.2487 \text{ g} = 0.249 \text{ g}$$

14.55 First, find the moles of nitric acid, HNO_3, using the molarity. Since 35.9 mL is 0.0359 L, you get

$$0.0359 \text{ L solution} \times \dfrac{0.0876 \text{ mol } HNO_3}{1 \text{ L solution}} = 0.003145 \text{ mol } HNO_3$$

The molecular weight of sodium carbonate, Na_2CO_3, is 105.99 amu. Therefore, the mass of Na_2CO_3 required is

$$0.003145 \text{ mol } HNO_3 \times \dfrac{1 \text{ mol } Na_2CO_3}{2 \text{ mol } HNO_3} \times \dfrac{105.99 \text{ g } Na_2CO_3}{1 \text{ mol } Na_2CO_3} = 0.1667 \text{ g} = 0.167 \text{ g}$$

The molecular weight of NaNO$_3$ is 85.00 amu. Therefore, the mass of NaNO$_3$ produced is

$$0.003145 \text{ mol HNO}_3 \times \frac{2 \text{ mol NaNO}_3}{2 \text{ mol HNO}_3} \times \frac{85.00 \text{ g NaNO}_3}{1 \text{ mol NaNO}_3} = 0.2673 \text{ g} = 0.267 \text{ g}$$

14.56 First, find the moles of Pb(NO$_3$)$_2$ using the molarity. Since 154 mL is 0.154 L, you get

$$0.154 \text{ L solution} \times \frac{0.180 \text{ mol Pb(NO}_3)_2}{1 \text{ L solution}} = 0.02772 \text{ mol Pb(NO}_3)_2$$

The molecular weight of PbCrO$_4$ is 323.2 amu. Therefore, the mass of PbCrO$_4$ produced is

$$0.02772 \text{ mol Pb(NO}_3)_2 \times \frac{1 \text{ mol PbCrO}_4}{1 \text{ mol Pb(NO}_3)_2} \times \frac{323.2 \text{ g K}_2\text{CrO}_4}{1 \text{ mol PbCrO}_4} = 8.959 \text{ g} = 8.96 \text{ g}$$

The molecular weight of KNO$_3$ is 101.11 amu. Therefore, the mass of KNO$_3$ produced is

$$0.02772 \text{ mol Pb(NO}_3)_2 \times \frac{2 \text{ mol KNO}_3}{1 \text{ mol Pb(NO}_3)_2} \times \frac{101.11 \text{ g KNO}_3}{1 \text{ mol KNO}_3} = 5.606 \text{ g} = 5.61 \text{ g}$$

14.57 Molarity = $\dfrac{0.192 \text{ mol NaOH}}{1 \text{ L solution}} \times 0.0432 \text{ L solution} \times \dfrac{1 \text{ mol HC}_2\text{H}_3\text{O}_2}{1 \text{ mol NaOH}}$

$$\times \frac{1}{0.0100 \text{ L solution}} = 0.8294 \text{ M} = 0.829 \text{ M}$$

14.58 Molarity = $\dfrac{0.248 \text{ mol NaOH}}{1 \text{ L solution}} \times 0.0242 \text{ L solution} \times \dfrac{1 \text{ mol HCl}}{1 \text{ mol NaOH}}$

$$\times \frac{1}{0.00100 \text{ L solution}} = 6.002 \text{ M} = 6.00 \text{ M}$$

14.59 First, the moles of HNO_3 in the dilute solution are

$$0.150 \text{ L solution} \times \frac{0.15 \text{ mol } HNO_3}{1 \text{ L solution}} = 0.0225 \text{ mol } HNO_3$$

The volume of concentrated acid is

$$0.0225 \text{ mol } HNO_3 \times \frac{1 \text{ L solution}}{15.9 \text{ mol } HNO_3} \times \frac{1000 \text{ mL}}{1 \text{ L}} = 1.42 \text{ mL} = 1.4 \text{ mL}$$

14.60 Substitute into the dilution equation. The volume of concentrated ammonia is

$$\text{Volume} = \frac{1.5 \text{ mol/L} \times 0.185 \text{ L}}{14.5 \text{ mol/L}} = 0.0191 \text{ L} = 0.019 \text{ L} = 19 \text{ mL}$$

14.61 Substitute into the dilution equation to obtain

$$2.75 \text{ mol/L} \times V_i = 0.80 \text{ mol/L} \times 0.405 \text{ L}$$

Solving for V_i gives

$$V_i = \frac{0.80 \text{ mol/L} \times 0.405 \text{ L}}{2.75 \text{ mol/L}} = 0.118 \text{ L} = 0.12 \text{ L}$$

Therefore, 0.12 L, or 1.2×10^2 mL of concentrated solution is required.

14.62 Substitute into the dilution equation to get

$$0.25 \text{ mol/L} \times V_i = 0.20 \text{ mol/L} \times 0.125 \text{ L}$$

Solving for V_i gives

$$V_i = \frac{0.20 \text{ mol/L} \times 0.125 \text{ L}}{0.25 \text{ mol/L}} = 0.100 \text{ L} = 0.10 \text{ L}$$

Therefore, 0.10 L, or 1.0×10^2 mL of concentrated solution is required.

14.63 Substitute into the dilution equation. The molarity of the solution is

$$\text{Molarity} = \frac{2.85 \text{ mol/L} \times 0.0500 \text{ L}}{0.250 \text{ L}} = 0.5700 \text{ M} = 0.570 \text{ M}$$

14.64 Substitute into the dilution equation. The molarity of the solution is

$$\text{Molarity} = \frac{12.1 \text{ mol/L} \times 0.0250 \text{ L}}{0.500 \text{ L}} = 0.60\underline{5}0 \text{ M} = 0.605 \text{ M}$$

14.65 The molecular weight of C_2H_5OH is 46.07 amu. The moles of C_2H_5OH are

$$2.56 \text{ g } C_2H_5OH \times \frac{1 \text{ mol } C_2H_5OH}{46.07 \text{ g } C_2H_5OH} = 0.05557 \text{ mol } C_2H_5OH$$

Since 38.6 g is 0.0386 kg, the molality is

$$\frac{0.05557 \text{ mol } C_2H_5OH}{0.0386 \text{ kg solvent}} = 1.4\underline{4}0 \text{ m} = 1.44 \text{ m}$$

14.66 The molecular weight of $C_{12}H_{22}O_{11}$ is 342.30 amu. The moles of sucrose are

$$4.81 \text{ g } C_{12}H_{22}O_{11} \times \frac{1 \text{ mol } C_{12}H_{22}O_{11}}{342.30 \text{ g } C_{12}H_{22}O_{11}} = 0.01405 \text{ mol } C_{12}H_{22}O_{11}$$

Since 120.0 g is 0.1200 kg, the molality is

$$\frac{0.01405 \text{ mol } C_{12}H_{22}O_{11}}{0.1200 \text{ kg solution}} = 0.11\underline{7}1 \text{ m} = 0.117 \text{ m}$$

14.67 Solve the molality relationship for kilograms of solvent:

$$\text{mass solvent} = \frac{\text{moles solute}}{\text{molality}} = \frac{0.10 \text{ mol } NH_2CONH_2}{0.34 \text{ mol/kg}} = 0.294 \text{ kg} = 294 \text{ g}$$

The mass of urea, NH_2CONH_2, in the solution is (molecular weight 60.06 amu):

$$0.10 \text{ mol } NH_2CONH_2 \times \frac{60.06 \text{ g } NH_2CONH_2}{1 \text{ mol } NH_2CONH_2} = 6.006 \text{ g } NH_2CONH_2$$

The total mass of the solution is 294 g + 6.006 g = 3$\underline{0}$0.006 g = 3.0×10^2 g.

SOLUTIONS ■ 209

14.68 Solve the molality relationship for kilograms of solvent:

$$\text{mass solvent} = \frac{0.25 \text{ mol } C_6H_{12}O_6}{0.47 \text{ mol/kg}} = 0.532 \text{ kg} = 532 \text{ g}$$

The mass of glucose, $C_6H_{12}O_6$, (molecular weight 180.01 amu) is

$$0.25 \text{ mol } C_6H_{12}O_6 \times \frac{180.16 \text{ g } C_6H_{12}O_6}{1 \text{ mol } C_6H_{12}O_6} = 45.04 \text{ g}$$

The total mass of the solution is 532 g + 45.04 g = 577.04 g = 5.8 x 10² g.

14.69 The freezing point depression is

$$\Delta T_f = K_f m = (1.86°C/m) \times (0.25 \text{ m}) = 0.465°C$$

Therefore, the freezing point of the solution is

0.00°C - 0.465°C = -0.465°C = -0.47°C

14.70 The freezing point depression is

$$\Delta T_f = K_f m = (1.86°C/m) \times (0.34 \text{ m}) = 0.632°C$$

Therefore, the freezing point of the solution is

0.00°C - 0.632°C = -0.632°C = -0.63°C

14.71 The boiling point elevation is

$$\Delta T_b = K_b m = (0.52°C/m) \times (0.25 \text{ m}) = 0.130°C$$

Therefore, the boiling point of the solution is

100.00°C + 0.130°C = 100.130°C = 100.13°C

14.72 The boiling point elevation is

$$\Delta T_b = K_b m = (0.52°C/m) \times (0.34\ m) = 0.177°C$$

Therefore, the boiling point of the solution is

$$100.00°C + 0.177°C = 100.1\underline{7}7°C = 100.18°C$$

14.73 Since KBr forms K^+ and Br^- ions in solution, the molality of the solution is doubled, or $2 \times 0.015\ m = 0.030\ m$. The freezing point depression is

$$\Delta T_f = K_f m = (1.86°C/m) \times (0.030\ m) = 0.0558°C$$

The freezing point of the solution is $0.000°C - 0.0558°C = -0.05\underline{5}8°C = -0.056°C$

The boiling point elevation is

$$\Delta T_b = K_b m = (0.52°C/m) \times (0.030\ m) = 0.0156°C$$

The boiling point of the solution is $100.000°C + 0.0156°C = 100.01\underline{5}6°C = 100.016°C$

14.74 Since Na_2SO_4 forms 2 Na^+ and one SO_4^{2-} ions in solution, the molality of the solution is tripled, or $3 \times 0.015\ m = 0.045\ m$. The freezing point depression is

$$\Delta T_f = K_f m = (1.86°C/m) \times (0.045\ m) = 0.0837°C$$

The freezing point of the solution is $0.000°C - 0.0837°C = -0.08\underline{3}7°C = -0.084°C$

The boiling point elevation is

$$\Delta T_b = K_b m = (0.52°C/m) \times (0.045\ m) = 0.0234°C$$

The boiling point of the solution is $100.000°C + 0.0234°C = 100.02\underline{3}4°C = 100.023°C$

14.75 The freezing point depression is

$$\Delta T_f = \text{f. p. pure solvent} - \text{f. p. solution} = 5.455°C - 4.886°C = 0.569°C$$

The molality of the solution is

$$\text{molality} = \frac{\Delta T_f}{K_f} = \frac{0.569°C}{5.12°C/m} = 0.111\ m$$

Since 112 g is 0.112 kg, the moles of cholesterol are

$$0.112\ \text{kg solvent} \times \frac{0.111\ \text{mol cholesterol}}{1\ \text{kg solvent}} = 0.0124\ \text{mol cholesterol}$$

Thus, 4.82 g of cholesterol = 0.0124 mol cholesterol, or

$$\text{Molar mass} = \frac{4.82\ g}{0.0124\ \text{mol}} = 387.2\ g/mol = 387\ g/mol,\ \text{or 387 amu}$$

14.76 The freezing point depression is

$$\Delta T_f = \text{f. p. pure solvent} - \text{f. p. solution} = 5.455°C - 4.850°C = 0.605°C$$

The molality of the solution is

$$\text{molality} = \frac{\Delta T_f}{K_f} = \frac{0.605°C}{5.12°C/m} = 0.1182\ m$$

Since 1.048 g = 1.048×10^{-3} kg, the moles of Vitamin K_1 are

$$1.048 \times 10^{-3}\ \text{kg solvent} \times \frac{0.1182\ \text{mol vitamin } K_1}{1\ \text{kg solvent}} = 1.238 \times 10^{-4}\ \text{mol vitamin } K_1$$

Thus, 55.8 mg, or 0.0558 g of Vitamin K_1 = 1.238×10^{-4} mol, or

$$\text{Molar mass} = \frac{0.0558\ g}{1.238 \times 10^{-4}\ \text{mol}} = 450.6\ g/mol = 451\ g/mol,\ \text{or 451 amu}$$

Solutions to Additional Problems

Note on significant figures: The first time the final answer is written, it is written first with one nonsignificant figure. The least significant digit is also underlined. The final answer is then rounded to the correct number of significant figures but the least significant digit is no longer underlined.

14.77 In exactly 100 g of this solution, the mass of solute is

$$\text{mass NH}_4\text{Cl} = \frac{9.30\% \times 100 \text{ g}}{100\%} = 9.30 \text{ g NH}_4\text{Cl}$$

The molecular weight of NH_4Cl is 53.49 amu. Thus

$$9.30 \text{ g NH}_4\text{Cl} \times \frac{1 \text{ mol NH}_4\text{Cl}}{53.49 \text{ g NH}_4\text{Cl}} = 0.1739 \text{ mol NH}_4\text{Cl}$$

The mass of water in the solution is

$$\text{mass H}_2\text{O} = \text{mass solution} - \text{mass NH}_4\text{Cl} = 100 \text{ g} - 9.30 \text{ g} = 90.70 \text{ g}$$

This corresponds to 0.09070 kg. The molality of the solution is

$$\text{molality} = \frac{\text{mol NH}_4\text{Cl}}{\text{kg H}_2\text{O}} = \frac{0.1739 \text{ mol}}{0.09070 \text{ kg}} = 1.9\underline{1}7 \text{ m} = 1.92 \text{ m}$$

Since the density of the solution is 1.04 g/mL, the volume of the solution is

$$100 \text{ g} \times \frac{1 \text{ mL}}{1.04 \text{ g}} = 96.15 \text{ mL} = 0.09615 \text{ L}$$

Thus, the molarity of the solution is

$$\text{molarity} = \frac{\text{mol NH}_4\text{Cl}}{\text{L solution}} = \frac{0.1739 \text{ mol}}{0.09615 \text{ L}} = 1.8\underline{0}9 \text{ M} = 1.81 \text{ M}$$

14.78 In exactly 100 g of this solution, the mass of solute is

$$\text{mass CaCl}_2 = \frac{6.25\% \times 100 \text{ g}}{100\%} = 6.25 \text{ g CaCl}_2$$

The molecular weight of $CaCl_2$ is 110.98 amu. Thus

$$6.25 \text{ g } CaCl_2 \times \frac{1 \text{ mol } CaCl_2}{110.98 \text{ g } CaCl_2} = 0.05632 \text{ mol } CaCl_2$$

The mass of water in the solution is

$$\text{mass } H_2O = \text{mass solution} - \text{mass } CaCl_2 = 100 \text{ g} - 6.25 \text{ g} = 93.75 \text{ g}$$

This corresponds to 0.09375 kg. Therefore, the molality of the solution is

$$\text{molality} = \frac{\text{mol } CaCl_2}{\text{kg } H_2O} = \frac{0.05632 \text{ mol}}{0.09375 \text{ kg}} = 0.6008 \text{ m} = 0.601 \text{ m}$$

Since the density of the solution is 1.03 g/mL, the volume of the solution is

$$100 \text{ g} \times \frac{1 \text{ mL}}{1.03 \text{ g}} = 97.09 \text{ mL} = 0.09709 \text{ L}$$

Therefore, the molarity of the solution is

$$\text{molarity} = \frac{\text{mol } CaCl_2}{\text{L solution}} = \frac{0.05632 \text{ mol}}{0.09709 \text{ L}} = 0.5801 \text{ M} = 0.580 \text{ M}$$

14.79 In two aqueous solutions, the lower freezing point will correspond to the solution with the higher effective molality. Since $BaCl_2$ dissolves in water to form one Ba^{2+} ion and two Cl^- ions, the effective molality is 3 x 0.10 m = 0.30 m. KCl dissolves in water to form one K^+ ion and one Cl^- ion. The effective molality is 2 x 0.10 m = 0.20 m. Since the $BaCl_2$ solution has the bigger effective molality, it will have the lower freezing point.

14.80 In two aqueous solutions, the lower freezing point will correspond to the solution with the higher effective molality. Since NaCl dissolves in water to form one Na^+ ion and one Cl^- ion, the effective molality is 2 x 0.22 m = 0.44 m. $AlCl_3$ dissolves in water to form one Al^{3+} ion and three Cl^- ions. The effective molality is 4 x 0.22 m = 0.88 m. Since the $AlCl_3$ solution has the bigger effective molality, it will have the lower freezing point.

214 ■ CHAPTER 14

14.81 Solve the freezing point depression equation for the molality. K_f for water is 1.86 °C/m, and ΔT_f is 2.78°C. Therefore,

$$\text{molality} = \frac{\Delta T_f}{K_f} = \frac{2.78\ °C}{1.86\ °C/m} = 1.4946\ m$$

This is the molality of ions in the solution. Since NaOCl dissociates into two ions, the molality of NaOCl in the solution is ½(1.4946 m) = 0.7473 m. In one kilogram of solution (1000 g), there are 0.7473 moles of solute. The molecular weight of NaOCl is 74.45 amu. Thus, the mass of NaOCl in the solution is

$$0.7473\ \text{mol} \times \frac{74.45\ g}{1\ \text{mol NaOCl}} = 55.64\ g\ \text{NaOCl}$$

The mass percent of NaOCl is

$$\%\ \text{mass} = \frac{55.64\ g\ \text{NaOCl}}{1000\ g\ \text{solution}} \times 100\ \% = 5.564\ \% = 5.56\ \%$$

14.82 Solve the freezing point depression equation for the molality. K_f for water is 1.86 °C/m, and ΔT_f is 1.63°C. Therefore,

$$\text{molality} = \frac{\Delta T_f}{K_f} = \frac{1.63\ °C}{1.86\ °C/m} = 0.8763\ m = 0.876\ m$$

In one kilogram of solvent (1000 g), there are 0.876 moles of solute. The molecular weight of H_2O_2 is 34.02 amu. Thus, the mass of H_2O_2 in the solution is

$$0.8763\ \text{mol} \times \frac{34.02\ g}{1\ \text{mol}\ H_2O_2} = 29.81\ g\ H_2O_2$$

The total mass of the solution is 1000 g + 29.81 g = 1029.81 g. The mass percent of H_2O_2 is

$$\%\ \text{mass} = \frac{29.81\ g\ H_2O_2}{1029.81\ g\ \text{solution}} \times 100\ \% = 2.895\ \% = 2.89\ \%$$

14.83 First, use the freezing point depression equation to calculate the molality of the solution. K_f for water is 1.86 °C/m, and ΔT_f is 4.36°C. Therefore,

$$\text{molality} = \frac{\Delta T_f}{K_f} = \frac{4.36\ °C}{1.86\ °C/m} = 2.344\ m$$

SOLUTIONS ■ 215

Next, use the boiling point elevation equation to calculate ΔT_b. K_b for water is 0.52 °C/m. Thus,

$$\Delta T_b = K_b m = (0.52 \,°C/m) \times (2.344 \,m) = 1.22 \,°C$$

Therefore, the boiling point of the solution is

boiling point = 100.00°C + 1.22°C = 101.22°C = 101.2°C.

14.84 First, use the boiling point elevation equation to calculate the molality of the solution. K_b for water is 0.52 °C/m, and ΔT_b = 102.4°C - 100.0°C = 2.4°C. Thus,

$$\text{molality} = \frac{\Delta T_b}{K_b} = \frac{2.4 \,°C}{0.52 \,°C/m} = 4.62 \,m$$

Next, use the freezing point depression equation to determine ΔT_f. K_f for water is 1.86 °C/m. Thus,

$$\Delta T_f = K_f m = (1.86 \,°C/m) \times (4.62 \,m) = 8.59 \,°C$$

Therefore, the solution will have a freezing point of

freezing point = 0.00°C - 8.59°C = -8.59°C = -8.6°C

14.85 The molarity of the phosphoric acid solution is

$$\text{molarity} = \frac{0.0152 \text{ mol Ba(OH)}_2}{1 \text{ L solution}} \times 0.0347 \text{ L solution} \times \frac{2 \text{ mol H}_3PO_4}{3 \text{ mol Ba(OH)}_2}$$

$$\times \frac{1}{0.00250 \text{ L solution}} = 0.1407 \,M = 0.141 \,M$$

The molecular weight of $Ba_3(PO_4)_2$ is 601.93 amu. The mass of the compound that is formed is

$$\text{mass } Ba_3(PO_4)_2 = \frac{0.0152 \text{ mol Ba(OH)}_2}{1 \text{ L solution}} \times 0.0347 \text{ L solution} \times \frac{1 \text{ mol Ba}_3(PO_4)_2}{3 \text{ mol Ba(OH)}_2}$$

$$\times \frac{601.93 \text{ g}}{1 \text{ mol Ba}_3(PO_4)_2} = 0.1058 \text{ g} = 0.106 \text{ g}$$

14.86 The molarity of the NaCl solution is

$$\text{molarity} = \frac{0.102 \text{ mol AgNO}_3}{1 \text{ L solution}} \times 0.0371 \text{ L solution} \times \frac{1 \text{ mol NaCl}}{1 \text{ mol AgNO}_3}$$

$$\times \frac{1}{0.0250 \text{ L solution}} = 0.1514 \text{ M} = 0.151 \text{ M}$$

The molecular weight of AgCl is 143.32 amu. The mass of the compound that forms is

$$\text{mass AgCl} = \frac{0.102 \text{ mol AgNO}_3}{1 \text{ L solution}} \times 0.0371 \text{ L solution} \times \frac{1 \text{ mol AgCl}}{1 \text{ mol AgNO}_3}$$

$$\times \frac{143.32 \text{ g}}{1 \text{ mol AgCl}} = 0.5424 \text{ g} = 0.542 \text{ g}$$

14.87 The amount of the HCl solution that must be added is

$$\text{volume HCl} = \frac{0.0145 \text{ mol Ca(OH)}_2}{1 \text{ L solution}} \times 1.00 \text{ L solution} \times \frac{2 \text{ mol HCl}}{1 \text{ mol Ca(OH)}_2}$$

$$\times \frac{1 \text{ L solution}}{2.50 \text{ mol HCl}} = 0.01160 \text{ L} = 0.0116 \text{ L}$$

Thus, 0.0116 L, or 11.6 mL of the HCl solution must be added.

14.88 The amount of the Ca(OH)$_2$ solution that must be added is

$$\text{volume Ca(OH)}_2 = \frac{0.00102 \text{ mol H}_2\text{SO}_4}{1 \text{ L solution}} \times 5.00 \text{ L solution} \times \frac{1 \text{ mol Ca(OH)}_2}{1 \text{ mol H}_2\text{SO}_4}$$

$$\times \frac{1 \text{ L solution}}{0.0208 \text{ mol Ca(OH)}_2} = 0.2452 \text{ L} = 0.245 \text{ L}$$

Thus, 0.245 L, or 245 mL of the Ca(OH)$_2$ solution must be added.

SOLUTIONS 217

14.89 First, determine the total number of moles of HCl that are in the final solution. For the 3.00 M solution, the moles of HCl are

$$\text{mol HCl} = \frac{3.00 \text{ mol HCl}}{1 \text{ L solution}} \times 0.148 \text{ L solution} = 0.4440 \text{ mol}$$

For the 6.25 M solution, the moles of HCl are

$$\text{mol HCl} = \frac{6.25 \text{ mol HCl}}{1 \text{ L solution}} \times 0.252 \text{ L solution} = 1.575 \text{ mol}$$

The total moles of HCl are 0.4440 mol + 1.575 mol = 2.019 mol. The total volume of the final solution is 148 mL + 252 mL = 400. mL, or 0.400 L. The molarity of the mixture is

$$\text{molarity} = \frac{2.019 \text{ mol}}{0.400 \text{ L}} = 5.048 \text{ M} = 5.05 \text{ M}$$

14.90 First, determine the total moles of NaOH that are in the final solution. For the 0.250 M solution, the moles of NaOH are

$$\text{mol NaOH} = \frac{0.250 \text{ mol NaOH}}{1 \text{ L solution}} \times 2.05 \text{ L solution} = 0.5125 \text{ mol}$$

For the 0.120 M solution, the moles of NaOH are

$$\text{mol NaOH} = \frac{0.120 \text{ mol NaOH}}{1 \text{ L solution}} \times 1.25 \text{ L solution} = 0.1500 \text{ mol}$$

The total moles of NaOH are 0.5125 mol + 0.1500 mol = 0.6625 mol. The total volume of the final solution is 2.05 L + 1.25 L = 3.30 L. Thus, the molarity of the mixture is

$$\text{molarity} = \frac{0.6625 \text{ mol}}{3.30 \text{ L}} = 0.2008 \text{ M} = 0.201 \text{ M}$$

14.91 The molarity of the solution can be determined as follows. In exactly 100 g of the solution, there are 10. g of Cu and 90. g of Ag. Silver is present in the largest amount, so it is the solvent. Thus, there is 0.090 kg of solvent in the solution. Next, find the moles of copper, atomic weight 63.55 amu. Thus,

$$\text{mol Cu} = 10. \text{ g} \times \frac{1 \text{ mol Cu}}{63.55 \text{ g}} = 0.157 \text{ mol}$$

The molality of the solution is

$$\text{molality} = \frac{0.157 \text{ mol}}{0.090 \text{ kg}} = 1.75 \text{ m}$$

ΔT_f for this solution is 961°C - 890°C = 71°C. Therefore, the freezing point depression constant for silver is

$$K_f = \frac{\Delta T_f}{m} = \frac{71 \text{ °C}}{1.75 \text{ m}} = 40.6 \text{ °C/m} = 41 \text{ °C/m}$$

14.92 The molarity of the solution can be determined as follows. In exactly 100 g of the solution, there are 33 g of Zn and 67 g of Cu. Copper is present in the largest amount, so it is the solvent. Thus, there is 0.067 kg of solvent in the solution. Next, find the moles of zinc, atomic weight 65.38 amu. Thus,

$$\text{mol Zn} = 33 \text{ g} \times \frac{1 \text{ mol Zn}}{65.38 \text{ g}} = 0.505 \text{ mol}$$

The molality of the solution is

$$\text{molality} = \frac{0.505 \text{ mol}}{0.067 \text{ kg}} = 7.54 \text{ m}$$

ΔT_f for this solution is 1083°C - 940°C = 143°C. Thus, the freezing point depression constant for copper is

$$K_f = \frac{\Delta T_f}{m} = \frac{143 \text{ °C}}{7.54 \text{ m}} = 19.0 \text{ °C/m} = 19 \text{ °C/m}$$

15. REACTION RATES AND CHEMICAL EQUILIBRIUM

■ Solutions to Exercises

Note on significant figures: The first time the final answer is written, it is written first with one nonsignificant figure. The least significant digit is also underlined. The final answer is then rounded to the correct number of significant figures but the least significant digit is no longer underlined.

15.1 The balanced equation is

$$4\ NH_3(g) + 3\ O_2(g) \rightleftharpoons 2\ N_2(g) + 6\ H_2O(g)$$

The equilibrium expression is

$$K = \frac{[N_2]^2\ [H_2O]^6}{[NH_3]^4\ [O_2]^3}$$

15.2 The balanced equation is

$$CH_4(g) + 2\ H_2S(g) \rightleftharpoons CS_2(g) + 4\ H_2(g)$$

The equilibrium expression is

$$K = \frac{[CS_2]\ [H_2]^4}{[CH_4]\ [H_2S]^2} = \frac{(1.10)\ (1.68)^4}{(1.10)\ (1.49)^2} = 3.5\underline{8}8 = 3.59$$

15.3 The equilibrium expression for the reaction is

$$K = \frac{[NO]^2}{[N_2][O_2]} = 4.6 \times 10^{-31}$$

The magnitude of the equilibrium constant is very small so the left side of the equation is favored. Therefore, the reaction will not go to completion.

15.4 The balanced chemical equation is

$$3\ Fe(s) + 4\ H_2O(s) \rightleftharpoons Fe_3O_4(s) + 4\ H_2(g)$$

Since iron (Fe) and iron oxide (Fe_3O_4) are pure solids, a concentration term for these substances does not appear in the equilibrium expression. Thus,

$$K = \frac{[H_2]^4}{[H_2O]^4}$$

15.5 Both $Ca(OH)_2$ and $Ba(OH)_2$ produce the same number of ions when they dissolve. Since K_{sp} for $Ba(OH)_2$ is larger than K_{sp} for $Ca(OH)_2$, $Ba(OH)_2$ is more soluble.

15.6 Since the concentration of H_2 is increased, the reaction will shift from left to right (away from H_2) in an attempt to remove some of the extra hydrogen.

15.7 Since the volume decreases, the pressure will increase and the equilibrium will shift towards fewer molecules. Thus, the equilibrium will shift from left to right.

15.8 Decreasing the temperature will cause the reaction to shift from right to left, towards the heat. Thus, the decomposition of carbon dioxide will decrease.

■ Answers to Questions to Test Your Reading

15.1 The primary factor that influences the rate of a reaction is the identity and characteristics of the reactants, and the secondary factors are the conditions, such as the concentrations of the reactants and the temperature of the reaction, under which the reaction occurs.

15.2 In order to react with each other, two molecules must collide.

15.3 According to collision theory, a collision between two molecules does not always result in a reaction. There is a minimum energy, the activation energy, that is required before reaction can occur. Not every collision will supply enough energy to break the chemical bonds necessary to cause a reaction.

15.4 If the concentration of a reactant is increased, this usually results in an increase in the rate of the reaction.

15.5 Reaction rates increase when the concentrations of the reactants are increased. This is because more molecules will exist in a given volume and thus, more collisions will occur.

15.6 Reaction rates increase with increasing temperature. This is because molecular energies and speeds increase, and thus the average collisional energy will also increase. The likelihood that a given collision will exceed the activation energy will also increase.

15.7 A rule of thumb: the rate of the reaction will double for each 10°C increase in the temperature.

15.8 A catalyst is a substance that increases the rate of a reaction by providing a new pathway for the reaction to occur. The new pathway will have a lower activation energy than the uncatalyzed pathway, and thus the reaction will have a faster rate.

15.9 The activation energy of an uncatalyzed reaction is higher than the activation energy of the same reaction in the presence of a catalyst. The catalyst lowers the activation energy.

15.10 The catalytic converter is mounted in the exhaust system and typically contains a platinum and rhodium catalyst. These catalysts allow the usually slow reactions that destroy pollutants, such as CO and NO, to occur more rapidly so that the pollutants are used up before they can leave the car.

15.11 Chemical Equilibrium is a dynamic state in which the rates of the forward and reverse reactions have become equal. A chemical system at equilibrium is not a static one because reactants are still being transformed into products and products are still reacting to become the original reactants.

15.12 The equilibrium constant is the value obtained when equilibrium concentrations are substituted into the equilibrium ratio, K. For the reaction

$$a A + b B \rightleftharpoons c C + d D$$

the equilibrium constant is

$$K = \frac{[C]^c [D]^d}{[A]^a [B]^b}$$

The equilibrium constant is a constant at a given temperature. If the temperature changes, so does K. Therefore, it is not always constant. (See Table 15.1)

15.13 Each reaction has its own characteristic equilibrium constant at a given temperature. However, there are an infinite number of equilibrium compositions, each obtained from different initial conditions, that correspond to the same equilibrium constant, K. (See Table 15.1)

15.14 The magnitude of the equilibrium constant tells us if a particular reaction favors the left or the right sides of the chemical equation. Large values of K, the reaction will favor the right side, and small values of K favor the left side. When the equilibrium constant is neither large nor small (around 1), neither side is favored.

15.15 Homogeneous equilibria are reactions in which all substances are in a single state of matter. Heterogeneous equilibria are reactions in which more than one state of matter is involved.

15.16 a. This reaction has two phases, solid, and aqueous. Thus, it is a heterogeneous equilibrium.

b. This reaction has only one phase, gas. Thus, it is a homogeneous equilibrium.

15.17 Le Chatelier's Principle states that when a system in chemical equilibrium is subjected to a change of a concentration, volume, or temperature, the equilibrium composition will shift in a way that tends to counteract the change.

15.18 You can alter the equilibrium composition of a reaction mixture in the following ways:
i. Changing the concentrations by adding or removing substances that appear on either side of the chemical equation (and in the equilibrium expression).
ii. Changing the volume of the system.
iii. Changing the temperature.

15.19 An exothermic reaction liberates heat. An endothermic reaction absorbs heat from the surroundings.

15.20 For an exothermic reaction at equilibrium, when the temperature is increased, the reaction will shift from right to left in an attempt to absorb the heat and counteract the temperature increase.

15.21 For an endothermic reaction at equilibrium, when the temperature is increased, the reaction will shift from left to right in an attempt to absorb the heat and counteract the temperature increase.

15.22 A catalyst has no effect on the equilibrium composition of a reaction; a catalyst merely speeds up the rate at which equilibrium is reached.

15.23 Platinum serves as a catalyst in the reaction. It increases the rate of the reaction. It has no effect on the equilibrium composition of the reaction mixture.

15.24 Four ways to increase the yield of ammonia are
  i. Add N_2.
  ii. Remove NH_3.
  iii. Decrease the volume.
  iv. Decrease the temperature.

■ Solutions to Practice Problems

Note on significant figures: The first time the final answer is written, it is written first with one nonsignificant figure. The least significant digit is also underlined. The final answer is then rounded to the correct number of significant figures but the least significant digit is no longer underlined.

15.25a. $2\ SO_2(g) + O_2(g) \rightleftharpoons 2\ SO_3(g)$ $\qquad K = \dfrac{[SO_3]^2}{[SO_2]^2\,[O_2]}$

  b. $2\ NOCl(g) \rightleftharpoons 2\ NO(g) + Cl_2(g)$ $\qquad K = \dfrac{[NO]^2\,[Cl_2]}{[NOCl]^2}$

  c. $POCl_3(g) \rightleftharpoons POCl(g) + Cl_2(g)$ $\qquad K = \dfrac{[POCl]\,[Cl_2]}{[POCl_3]}$

  d. $2\ CO(g) + O_2(g) \rightleftharpoons 2\ CO_2(g)$ $\qquad K = \dfrac{[CO_2]^2}{[CO]^2\,[O_2]}$

15.26a. $2\ H_2O(g) \rightleftharpoons 2\ H_2(g) + O_2(g)$ $\qquad K = \dfrac{[H_2]^2\,[O_2]}{[H_2O]^2}$

  b. $2\ NO(g) + O_2(g) \rightleftharpoons 2\ NO_2(g)$ $\qquad K = \dfrac{[NO_2]^2}{[NO]^2\,[O_2]}$

  c. $CO(g) + 2\ H_2(g) \rightleftharpoons CH_3OH(g)$ $\qquad K = \dfrac{[CH_3OH]}{[CO]\,[H_2]^2}$

  d. $CO(g) + H_2O(g) \rightleftharpoons CO_2(g) + H_2(g)$ $\qquad K = \dfrac{[CO_2]\,[H_2]}{[CO]\,[H_2O]}$

15.27a. $CS_2(g) + 4 H_2(g) \rightleftharpoons CH_4(g) + 2 H_2S(g)$ $K = \dfrac{[CH_4][H_2S]^2}{[CS_2][H_2]^4}$

b. $I_2(g) + Br_2(g) \rightleftharpoons 2 IBr(g)$ $K = \dfrac{[IBr]^2}{[I_2][Br_2]}$

c. $COCl_2(g) \rightleftharpoons CO(g) + Cl_2(g)$ $K = \dfrac{[CO][Cl_2]}{[COCl_2]}$

d. $NH_4^+(aq) \rightleftharpoons NH_3(aq) + H^+(aq)$ $K = \dfrac{[NH_3][H^+]}{[NH_4^+]}$

15.28a. $2 NO_2(g) + 7 H_2(g) \rightleftharpoons 2 NH_3(g) + 4 H_2O(g)$ $K = \dfrac{[NH_3]^2[H_2O]^4}{[NO_2]^2[H_2]^7}$

b. $IF(g) + F_2(g) \rightleftharpoons IF_3(g)$ $K = \dfrac{[IF_3]}{[IF][F_2]}$

c. $4 HCl(g) + O_2(g) \rightleftharpoons 2 Cl_2(g) + 2 H_2O(g)$ $K = \dfrac{[Cl_2]^2[H_2O]^2}{[HCl]^4[O_2]}$

d. $HCN(aq) \rightleftharpoons H^+(aq) + CN^-(aq)$ $K = \dfrac{[H^+][CN^-]}{[HCN]}$

15.29 $CO(g) + 3 H_2(g) \rightleftharpoons CH_4(g) + H_2O(g)$

$K = \dfrac{[CH_4][H_2O]}{[CO][H_2]^3} = \dfrac{(0.0387)(0.0387)}{(0.0613)(0.184)^3} = 3.9\underline{22} = 3.92$

15.30 The balanced equation is

$$2\ NO(g) + 2\ H_2(g) \rightleftharpoons N_2(g) + 2\ H_2O(g)$$

The equilibrium constant is

$$K = \frac{[N_2][H_2O]^2}{[NO]^2[H_2]^2} = \frac{(0.019)(0.138)^2}{(0.062)^2(0.012)^2} = 654 = 6.5 \times 10^2$$

15.31 $H_2(g) + I_2(g) \rightleftharpoons 2\ HI(g)$

$$K = \frac{[HI]^2}{[H_2][I_2]} = \frac{(0.0157)^2}{(0.00213)(0.00213)} = 54.33 = 54.3$$

15.32 $2\ SO_2(g) + O_2(g) \rightleftharpoons 2\ SO_3(g)$

$$K = \frac{[SO_3]^2}{[SO_2]^2[O_2]} = \frac{(0.260)^2}{(0.590)^2(0.0450)} = 4.315 = 4.32$$

15.33 The equilibrium expression is

$$K = \frac{[H_2O]^2}{[H_2]^2[O_2]} = 3 \times 10^{81}$$

The magnitude of the equilibrium constant is very large, so the right side of the equation is favored. Therefore, the reaction will go to completion.

15.34 The equilibrium expression is

$$K = \frac{[NO_2]^2}{[N_2]^2[O_2]} = 3 \times 10^{-17}$$

The magnitude of the equilibrium constant is very small, so the left side of the equation is favored. Thus, we do not expect a complete reaction.

15.35a. $C(s) + CO_2(g) \rightleftharpoons 2\ CO(g)$ $K = \dfrac{[CO]^2}{[CO_2]}$

  b. $FeO(s) + CO(g) \rightleftharpoons Fe(s) + CO_2(g)$ $K = \dfrac{[CO_2]}{[CO]}$

c. $2\ Na_2CO_3(s) + 2\ SO_2(g) + O_2(g) \rightleftharpoons 2\ Na_2SO_4(s) + 2\ CO_2(g)$

$$K = \frac{[CO_2]^2}{[SO_2]^2\,[O_2]}$$

15.36a. $NH_4Cl(s) \rightleftharpoons NH_3(g) + HCl(g)$ $K = [NH_3]\,[HCl]$

b. $C(s) + 2\ N_2O(g) \rightleftharpoons CO_2(g) + 2\ N_2(g)$ $K = \dfrac{[CO_2]\,[N_2]^2}{[N_2O]^2}$

c. $2\ NaHCO_3(s) \rightleftharpoons Na_2CO_3(s) + H_2O(g) + CO_2(g)$ $K = [H_2O]\,[CO_2]$

15.37a. $P_4(s) + 5\ O_2(g) \rightleftharpoons P_4O_{10}(s)$ $K = \dfrac{1}{[O_2]^5}$

b. $2\ H_2O_2(l) \rightleftharpoons 2\ H_2O(l) + O_2(g)$ $K = [O_2]$

c. $PbI_2(s) + Cl_2(g) \rightleftharpoons PbCl_2(s) + I_2(g)$ $K = \dfrac{[I_2]}{[Cl_2]}$

15.38a. $4\ CuO(s) \rightleftharpoons 2\ Cu_2O(s) + O_2(g)$ $K = [O_2]$

b. $2\ NaBr(s) + Cl_2(g) \rightleftharpoons 2\ NaCl(s) + Br_2(l)$ $K = \dfrac{[Br_2]}{[Cl_2]}$

c. $2\ Cr(s) + 3\ Cl_2(g) \rightleftharpoons 2\ CrCl_3(s)$ $K = \dfrac{1}{[Cl_2]^3}$

15.39 Both PbI_2 and Ag_2CrO_4 produce the same number of ions when they dissolve. Since K_{sp} for PbI_2 is larger than K_{sp} for Ag_2CrO_4, PbI_2 is more soluble.

REACTION RATES AND CHEMICAL EQUILIBRIUM ■ 227

15.40 Both CaF_2 and Ag_2S produce the same number of ions when they dissolve. Since K_{sp} for CaF_2 is larger than K_{sp} for Ag_2S, CaF_2 is more soluble.

15.41 Since O_2 has been added, the reaction will shift from left to right (away from O_2) to remove some of the extra oxygen.

15.42 Since CH_4 has been added, the reaction will shift from right to left (away from CH_4) to remove some of the extra methane.

15.43 Since the volume decreases, the pressure increases and the equilibrium will shift towards fewer molecules. Thus, the equilibrium shifts from left to right.

15.44 Since the volume decreases, the pressure increases and the equilibrium will shift towards fewer molecules. However, both sides of the equation have the same number of molecules. Thus, the decrease in volume will not affect the equilibrium.

15.45 For an exothermic reaction, when the temperature is raised, the reaction shifts from right to left, to absorb the heat being added. Thus, CH_3OH would decompose into CO and H_2.

15.46 For an endothermic reaction, when the temperature is raised, the reaction shifts from left to right, to absorb the heat being added. Thus, more NO would form.

15.47 The equilibrium constant corresponds to the equation

$$CH_4 + 2\,H_2S \rightleftharpoons CS_2 + 4\,H_2$$

15.48 The equilibrium constant corresponds to the equation

$$4\,NO + 6\,H_2O \rightleftharpoons 4\,NH_3 + 5\,O_2$$

15.49 The equilibrium constant corresponds to the equation

$$2\,NO + 2\,H_2 \rightleftharpoons N_2 + 2\,H_2O$$

15.50 The equilibrium constant corresponds to the equation

$$4\,HCl + O_2 \rightleftharpoons 2\,H_2O + 2\,Cl_2$$

15.51 The equilibrium expression is

$$K = [Pb^{2+}][Br^-]^2 = (0.010)(0.020)^2 = 4.00 \times 10^{-6} = 4.0 \times 10^{-6}$$

228 ■ CHAPTER 15

15.52 The equilibrium expression is

$$K = [Ag^+]^3 [PO_4^{3-}] = (4.8 \times 10^{-5})^3 (1.6 \times 10^{-5}) = 1.77 \times 10^{-18} = 1.8 \times 10^{-18}$$

■ Solutions to Additional Problems

15.53a. If H_2 is removed, the reaction will shift from right to left to replace the H_2 removed.

    b. If N_2 is added, the reaction will shift from right to left to replace the N_2 added.

    c. If the volume is decreased, the pressure is increased and the reaction will shift towards fewer molecules. Thus, the reaction shifts from left to right.

    d. If the temperature is decreased, for an exothermic reaction, the reaction shifts from left to right.

    e. If a catalyst is added, there will be no change in the equilibrium composition.

15.54a. If SO_3 is added, the reaction will shift from left to right to use up the added SO_3.

    b. If SO_2 is removed, the reaction will shift from left to right to replace the SO_2 removed.

    c. If the volume is decreased, the pressure is increased and the reaction will shift towards fewer molecules. Thus, the reaction shifts from right to left.

    d. For an endothermic reaction, when the temperature is decreased, the reaction shifts from left to right.

    e. A catalyst has no effect on the equilibrium composition.

15.55 For parts a, b, c, and e, there is no effect on K. For part d, K increases.

15.56 For parts a, b, c, and e, there is no effect on K. For part d, K increases.

15.57a. If Cl^- is removed, the reaction will shift from left to right. More $PbCl_2(s)$ will dissolve.

    b. If $Pb(NO_3)_2$ is added, the reaction will shift from right to left. $PbCl_2(s)$ will form.

    c. If $PbCl_2(s)$ is added to the saturated solution, no change will occur. The excess $PbCl_2$ will settle to the bottom of the vessel.

    d. For an endothermic process, when the temperature is increased, the reaction shifts from left to right.

15.58a. If $BaCO_3(s)$ is added, no change will occur. The excess $BaCO_3$ will settle to the bottom of the vessel.

b. If $BaO(s)$ is removed, no change will occur as long as some of the solid remains in the equilibrium mixture.

c. If CO_2 is added, the reaction will shift from left to right.

d. If the temperature is decreased, for an exothermic process, the reaction shifts from left to right.

15.59 For parts a, b, and c, there is no effect on K. For part d, K will increase.

15.60 For parts a, b, and c, there is no effect on K. For part d, K will increase.

16. ACIDS AND BASES

■ Solutions to Exercises

Note on significant figures: The first time the final answer is written, it is written first with one nonsignificant figure. The least significant digit is also underlined. The final answer is then rounded to the correct number of significant figures but the least significant digit is no longer underlined.

16.1 $PbSO_4$ is an insoluble ionic compound. The other compounds are soluble. The molecular equation is

$$Pb(NO_3)_2(aq) + Na_2SO_4(aq) \longrightarrow PbSO_4(s) + 2\ NaNO_3(aq)$$

Writing the formulas of the ionic compounds, you obtain the total ionic equation.

$$Pb^{2+}(aq) + 2\ NO_3^-(aq) + 2\ Na^+(aq) + SO_4^{2-}(aq) \longrightarrow$$

$$PbSO_4(s) + 2\ Na^+(aq) + 2\ NO_3^-(aq)$$

The sodium and nitrate ions are spectator ions because they have not undergone any change. Thus, they can be eliminated.

$$Pb^{2+}(aq) + 2\ \cancel{NO_3^-}(aq) + 2\ \cancel{Na^+}(aq) + SO_4^{2-}(aq) \longrightarrow$$

$$PbSO_4(s) + 2\ \cancel{Na^+}(aq) + 2\ \cancel{NO_3^-}(aq)$$

The net ionic equation is

$$Pb^{2+}(aq) + SO_4^{2-}(aq) \longrightarrow PbSO_4(s)$$

16.2 $HBr(g) + H_2O(l) \longrightarrow H_3O^+(aq) + Br^-(aq)$

16.3 The conjugate acid of OH^- can be obtained by adding an H^+ ion to it

$$OH^- \xrightarrow{+ H^+} H_2O$$

Therefore, H_2O is the conjugate acid of OH^-.

16.4 $H_2CO_3(aq) + CN^-(aq) \longrightarrow HCN(aq) + HCO_3^-(aq)$
  acid   base   acid   base

Thus, H_2CO_3 and HCO_3^- are a conjugate acid-base pair, as are CN^- and HCN.

16.5 With $[H_3O^+]$ = 8.4 x 10^{-5} M, obtain the concentration of hydroxide ions from the expression for K_W.

$K_W = [H_3O^+][OH^-]$

$1.0 \times 10^{-14} = (8.4 \times 10^{-5}) \times [OH^-]$

$[OH^-] = \dfrac{1.0 \times 10^{-14}}{8.4 \times 10^{-5}} = 1.19 \times 10^{-10}$ M = 1.2×10^{-10} M

16.6 Since 4.7 x 10^{-5} M is greater than 1.0 x 10^{-7} M, the solution is acidic.

16.7 On a calculator, do the following operations:

1. Press 2.2
2. Press EXP (or EE)
3. Press 11
4. Press +/-
5. Press LOG
6. Press +/-

The result is 10.6_5_8. Thus, pH = 10.66.

16.8 Do the following operations on your calculator.

1. Press 5
2. Press EXP (or EE)
3. Press 8
4. Press +/-
5. Press LOG
6. Press +/-

The result is 7.30 Thus, the pH = 7.3, and the solution is basic.

16.9 Using a calculator, enter the following operations

1. Press 3.16
2. Press +/-
3. Press INV
4. Press LOG

The result is 6.92×10^{-4}. Thus,

$$[H_3O^+] = 6.9 \times 10^{-4} \, M$$

The solution is acidic.

■ Answers to Questions to Test Your Reading

16.1 According to Arrhenius, <u>acids</u> are substances that produce H^+ ions when they are dissolved in water and <u>bases</u> produce OH^- ions. Examples are nitric acid, HNO_3, and sodium hydroxide, NaOH.

16.2 The net ionic equation that describes the reaction between HBr and KOH is

$$H^+ (aq) + OH^- (aq) \longrightarrow H_2O (l)$$

Br^- and K^+ are spectator ions.

16.3 The net ionic equation for both reactions is

$$H^+ (aq) + OH^- (aq) \longrightarrow H_2O (l)$$

All other ions in both reactions are spectator ions. Therefore, since the net ionic equation is the same for both reactions, the quantity of heat liberated is identical.

ACIDS AND BASES

16.4 According to the Bronsted-Lowry concept, an <u>acid</u> is a proton (H^+) donor and a <u>base</u> is a proton acceptor. Examples of an acid and a base are hydrofluoric acid (HF) and ammonia (NH_3).

16.5 Hydronium ion is H_3O^+ and has the structure

$$H - \overset{..}{\underset{H}{O}} - H \quad ^+$$

16.6 A <u>conjugate acid-base pair</u> consists of two substances (one an acid and the other a base) in an acid-base reaction that differ by the gain or loss of a proton. For example

$$\underset{\text{acid}}{HSO_4^-(aq)} + \underset{\text{base}}{CN^-(aq)} \longrightarrow \underset{\text{base}}{SO_4^{2-}(aq)} + \underset{\text{acid}}{HCN(aq)}$$

The HSO_4^- and the SO_4^{2-} ions are a conjugate acid-base pair, as are CN^- and HCN.

16.7 The reaction of an acid with water is represented by

$$\underset{\text{acid}}{HA(aq)} + \underset{\text{base}}{H_2O(l)} \rightleftharpoons \underset{\text{acid}}{H_3O^+(aq)} + \underset{\text{base}}{A^-(aq)}$$

16.8 Water is <u>amphoteric</u> because it can behave as both an acid and a base. This is best seen in the self-ionization of water in which a proton is transferred from one water molecule to another.

$$\underset{\text{acid}}{H_2O(l)} + \underset{\text{base}}{H_2O(l)} \rightleftharpoons \underset{\text{acid}}{H_3O^+(aq)} + \underset{\text{base}}{OH^-(aq)}$$

Thus, H_2O is the conjugate acid of OH^- and the conjugate base of H_3O^+.

16.9 A <u>monoprotic</u> acid can only transfer one proton to a base. An example is HCl, hydrochloric acid. A <u>diprotic</u> acid can transfer two protons. An example is H_2SO_4, sulfuric acid.

16.10 Acids with very large equilibrium constants are strong acids. They are completely dissociated in water. They are: HCl, hydrochloric acid; HBr, hydrobromic acid; HI, hydroiodic acid; HNO_3, nitric acid; $HClO_4$, perchloric acid; and H_2SO_4, sulfuric acid. (See Table 16.1)

16.11 Acids with small equilibrium constants are weak acids. They are not completely dissociated in water. An example is H_2SO_3, sulfurous acid. (See Table 16.2)

234 ■ CHAPTER 16

16.12 Since the strength of a conjugate base increases as the strength of an acid decreases, the strongest bases have the weakest conjugate acids. They are totally ionized in water. They are: LiOH, lithium hydroxide; NaOH, sodium hydroxide; KOH, potassium hydroxide; $Ca(OH)_2$, calcium hydroxide; $Sr(OH)_2$, strontium hydroxide; and $Ba(OH)_2$, barium hydroxide. (See Table 16.1)

16.13 A weak base is a base that is not totally ionized in water. An example is ammonia, NH_3.

16.14 An <u>acid-ionization constant (K_a)</u> is the equilibrium constant for the dissociation of an acid in water.

16.15 The equilibrium expression for the reaction of HNO_2 with water is

$$K_a = \frac{[H_3O^+][NO_2^-]}{[HNO_2]}$$

16.16 The stronger the acid, the weaker the conjugate base. Since formic acid, $HCHO_2$, is a stronger acid than acetic acid, $HC_2H_3O_2$, therefore formate ion, $HCHO_2^-$ is a weaker base than acetate ion, $C_2H_3O_2^-$.

16.17 Since formic acid is a stronger acid than acetic acid, it has a greater tendency to transfer protons to water. Thus, a 0.1M solution of formic acid will transfer more protons to water than a 0.1M solution of acetic acid, and thus have a higher concentration of hydronium ions.

16.18 H_2SO_3 is a stronger acid than HSO_3^- because HSO_3^- is the conjugate base of H_2SO_3.

16.19 <u>Self-ionization</u> means that two identical molecules react to give ions. The self-ionization of water to give H_3O^+ and OH^- ions is an example.

16.20 The <u>ion-product constant</u> for water (K_W) is

$$K_W = [H_3O^+][OH^-] = 1.0 \times 10^{-14} \text{ at } 25°C$$

16.21 In terms of hydrogen ion concentrations, in an acidic solution the concentration of H^+ ions is greater than that of OH^- ions. In a basic solution, the concentration of OH^- ions is greater than that of H^+ ions. In a neutral solution, the concentrations of H^+ and OH^- ions are equal. At 25°C,

In an acidic solution, $[H_3O^+] > 1.0 \times 10^{-7}$ M
In a neutral solution, $[H_3O^+] = 1.0 \times 10^{-7}$ M
In a basic solution, $[H_3O^+] < 1.0 \times 10^{-7}$ M

ACIDS AND BASES ■ 235

16.22 pH is defined as a unitless number obtained from the negative of the logarithm of the hydrogen ion concentrations.

$$pH = -\log[H_3O^+]$$

The pH scale normally runs from 0 to 14. However, for strong acids or bases, pH values below -1 and above 15, respectively, are possible.

In an acidic solution, pH < 7.00
In a neutral solution, pH = 7.00
In a basic solution, pH > 7.00

16.23 Rainwater with a pH of 4.3 would be acidic.

16.24 The shampoo with the lower pH is more acidic. Thus, the shampoo with pH of 6.5 is more acidic.

16.25 pH can be measured in the following two ways. First, using a pH meter. Second, using acid-base indicators.

16.26 An indicator is a weak acid whose solution will change color within a small pH range and indicate the pH of the solution by the color.

16.27 Bromocresol green is yellow below pH 3.8 and blue above pH 5.4. Thus, when a solution that contains bromocresol green is green, the pH is approximately 4-5.

16.28 When bromocresol green is yellow, the pH is less than 3.8. When methyl violet is violet, the pH is above 1.6. Thus the solution has a pH of approximately 2-4.

16.29 A <u>buffer</u> is a solution characterized by its ability to resist changes in pH when limited amounts of acid or base are added to it. They contain a weak acid and its conjugate base. An example is NH_4^+ and NH_3.

16.30 When a strong acid is added to an ammonia-ammonium chloride buffer, H_3O^+ ions will react with the base, NH_3, to form NH_4^+, the conjugate acid. When a strong base is added to the buffer, it reacts with the acid, NH_4^+, to form the conjugate base, NH_3.

Solutions to Practice Problems

Note on significant figures: The first time the final answer is written, it is written first with one nonsignificant figure. The least significant digit is also underlined. The final answer is then rounded to the correct number of significant figures but the least significant digit is no longer underlined.

16.31 Both AgOH and AgCl are insoluble in water. The total ionic equation for this reaction is

$$AgOH(s) + H^+(aq) + Cl^-(aq) \longrightarrow AgCl(s) + H_2O(l)$$

Since there are no spectator ions, this is also the net ionic equation.

16.32 $CaCO_3$ is insoluble in water. The other compounds are soluble. The total ionic equation is

$$Ca^{2+}(aq) + 2\,Cl^-(aq) + 2\,Li^+(aq) + CO_3^{2-}(aq) \longrightarrow$$

$$CaCO_3(s) + 2\,Li^+(aq) + 2\,Cl^-(aq)$$

Li^+ and Cl^- are spectator ions. The net ionic equation is

$$Ca^{2+}(aq) + CO_3^{2-}(aq) \longrightarrow CaCO_3(s)$$

16.33 $\quad HClO_4(l) + H_2O(l) \longrightarrow H_3O^+(aq) + ClO_4^-(aq)$

16.34 $\quad HI(g) + H_2O(l) \longrightarrow H_3O^+(aq) + I^-(aq)$

16.35 a. F^- b. HS^- c. S^{2-} d. OH^-

16.36 a. $C_2H_3O_2^-$ b. Cl^- c. ClO^- d. NO_2^-

16.37 a. NH_4^+ b. H_2S c. HS^- d. HCN

16.38 a. HF b. H_2SO_3 c. HSO_3^- d. $HClO_4$

ACIDS AND BASES

16.39a. $H_2SO_3(aq)$ + $H_2O(l)$ ⇌ $H_3O^+(aq)$ + $HSO_3^-(aq)$
       acid   base   acid   base

The conjugate acid-base pairs are H_2SO_3 with HSO_3^-, and H_2O with H_3O^+.

b. $HSO_3^-(aq)$ + $H_2O(l)$ ⇌ $H_3O^+(aq)$ + $SO_3^{2-}(aq)$
     acid   base   acid   base

The conjugate acid-base pairs are HSO_3^- with SO_3^{2-}, and H_2O with H_3O^+.

c. $CH_3NH_2(aq)$ + $H_2O(l)$ ⇌ $CH_3NH_3^+(aq)$ + $OH^-(aq)$
     base   acid   acid   base

The conjugate acid-base pairs are CH_3NH_2 with $CH_3NH_3^+$ and H_2O with OH^-.

d. $NH_4^+(aq)$ + $CN^-(aq)$ ⇌ $NH_3(aq)$ + $HCN(aq)$
     acid   base   base   acid

The conjugate acid-base pairs are NH_4^+ with NH_3 and CN^- with HCN.

16.40a. $H_2CO_3(aq)$ + $H_2O(l)$ ⇌ $H_3O^+(aq)$ + $HCO_3^-(aq)$
       acid   base   acid   base

The conjugate acid-base pairs are H_2CO_3 with HCO_3^- and H_2O with H_3O^+.

b. $HCO_3^-(aq)$ + $H_2O(l)$ ⇌ $H_3O^+(aq)$ + $CO_3^{2-}(aq)$
     acid   base   acid   base

The conjugate acid-base pairs are HCO_3^- with CO_3^{2-} and H_2O with H_3O^+.

c. $NH_3(aq) + H_2O(l) \rightleftharpoons NH_4^+(aq) + OH^-(aq)$
      base   acid   acid   base

The conjugate acid-base pairs are NH_3 with NH_4^+ and H_2O with OH^-.

d. $HPO_4^{2-}(aq) + NH_4^+(aq) \rightleftharpoons H_2PO_4^-(aq) + NH_3(aq)$
      base   acid   acid   base

The conjugate acid-base pairs are HPO_4^{2-} with $H_2PO_4^-$ and NH_4^+ with NH_3.

16.41 a. $CN^-(aq) + H_2O(l) \rightleftharpoons HCN(aq) + OH^-(aq)$
      base   acid   acid   base

The conjugate acid-base pairs are CN^- with HCN and H_2O with OH^-.

b. $HC_2H_3O_2(aq) + OH^-(aq) \rightleftharpoons C_2H_3O_2^-(aq) + H_2O(l)$
      acid   base   base   acid

The conjugate acid-base pairs are $HC_2H_3O_2$ with $C_2H_3O_2^-$ and OH^- with H_2O.

c. $HPO_4^{2-}(aq) + H_3O^+(aq) \rightleftharpoons H_2PO_4^-(aq) + H_2O(l)$
      base   acid   acid   base

The conjugate acid-base pairs are HPO_4^{2-} with $H_2PO_4^-$ and H_3O^+ with H_2O.

d. $HNO_2(aq) + H_2O(l) \rightleftharpoons NO_2^-(aq) + H_3O^+(aq)$
      acid   base   base   acid

The conjugate acid-base pairs are HNO_2 with NO_2^- and H_2O with H_3O^+.

16.42a. F⁻(aq) + H₂O(l) ⇌ HF(aq) + OH⁻(aq)
   base   acid   acid   base

The conjugate acid-base pairs are F⁻ with HF and H₂O with OH⁻.

b. F⁻(aq) + H₃O⁺(aq) ⇌ HF(aq) + H₂O(l)
   base   acid   acid   base

The conjugate acid-base pairs are F⁻ with HF and H₃O⁺ with H₂O.

c. H₂S(aq) + NH₃(aq) ⇌ NH₄⁺(aq) + HS⁻(aq)
   acid   base   acid   base

The conjugate acid-base pairs are H₂S with HS⁻ and NH₃ with NH₄⁺.

d. H₃O⁺(aq) + HS⁻(aq) ⇌ H₂S(aq) + H₂O(l)
   acid   base   acid   base

The conjugate acid-base pairs are H₃O⁺ with H₂O and HS⁻ with H₂S.

16.43a. $[OH^-] = \dfrac{K_w}{[H_3O^+]} = \dfrac{1.0 \times 10^{-14}}{0.25} = 4.00 \times 10^{-14}\ M = 4.0 \times 10^{-14}\ M$

b. $[OH^-] = \dfrac{K_w}{[H_3O^+]} = \dfrac{1.0 \times 10^{-14}}{0.0035} = 2.86 \times 10^{-12}\ M = 2.9 \times 10^{-12}\ M$

c. $[OH^-] = \dfrac{K_w}{[H_3O^+]} = \dfrac{1.0 \times 10^{-14}}{1.2 \times 10^{-2}} = 8.33 \times 10^{-13}\ M = 8.3 \times 10^{-13}\ M$

d. $[OH^-] = \dfrac{K_w}{[H_3O^+]} = \dfrac{1.0 \times 10^{-14}}{4.2 \times 10^{-11}} = 2.38 \times 10^{-4}\ M = 2.4 \times 10^{-4}\ M$

16.44a. $[H_3O^+] = \dfrac{K_w}{[OH^-]} = \dfrac{1.0 \times 10^{-14}}{0.15} = 6.67 \times 10^{-14}\ M = 6.7 \times 10^{-14}\ M$

b. $[H_3O^+] = \dfrac{K_w}{[OH^-]} = \dfrac{1.0 \times 10^{-14}}{3 \times 10^{-4}} = 3.3 \times 10^{-11}$ M $= 3 \times 10^{-11}$ M

c. $[H_3O^+] = \dfrac{K_w}{[OH^-]} = \dfrac{1.0 \times 10^{-14}}{5.3 \times 10^{-9}} = 1.89 \times 10^{-6}$ M $= 1.9 \times 10^{-6}$ M

d. $[H_3O^+] = \dfrac{K_w}{[OH^-]} = \dfrac{1.0 \times 10^{-14}}{0.00061} = 1.64 \times 10^{-11}$ M $= 1.6 \times 10^{-11}$ M

16.45 $[OH^-] = \dfrac{K_w}{[H_3O^+]} = \dfrac{1.0 \times 10^{-14}}{3.7 \times 10^{-4}} = 2.70 \times 10^{-11}$ M $= 2.7 \times 10^{-11}$ M

16.46 $[H_3O^+] = \dfrac{K_w}{[OH^-]} = \dfrac{1.0 \times 10^{-14}}{8.8 \times 10^{-4}} = 1.14 \times 10^{-11}$ M $= 1.1 \times 10^{-11}$ M

16.47 a. $[H_3O^+] = 5 \times 10^{-9}$ M, and is less than 1.0×10^{-7} M. Therefore, the solution is basic.

b. First, obtain the concentration of H_3O^+ ions using the expression for K_w.

$[H_3O^+] = \dfrac{K_w}{[OH^-]} = \dfrac{1.0 \times 10^{-14}}{1.0 \times 10^{-7}} = 1.00 \times 10^{-7}$ M $= 1.0 \times 10^{-7}$ M

Thus, the solution is neutral.

c. First, obtain the concentration of H_3O^+ ions using the expression for K_w.

$[H_3O^+] = \dfrac{K_w}{[OH^-]} = \dfrac{1.0 \times 10^{-14}}{5 \times 10^{-9}} = 2.0 \times 10^{-6}$ M $= 2 \times 10^{-6}$ M

Since 2×10^{-6} M is greater than 1.0×10^{-7} M, the solution is acidic.

d. $[H_3O^+] = 2 \times 10^{-7}$ M, and is greater than 1.0×10^{-7} M. Therefore, the solution is acidic.

ACIDS AND BASES ■ 241

16.48a. First, obtain the concentration of H_3O^+ ions using the expression for K_w.

$$[H_3O^+] = \frac{K_w}{[OH^-]} = \frac{1.0 \times 10^{-14}}{2 \times 10^{-4}} = 5.0 \times 10^{-11} \text{ M} = 5 \times 10^{-11} \text{ M}$$

Since 5×10^{-11} M is less than 1.0×10^{-7} M, the solution is basic.

b. First, obtain the concentration of H_3O^+ ions using the expression for K_w.

$$[H_3O^+] = \frac{K_w}{[OH^-]} = \frac{1.0 \times 10^{-14}}{6 \times 10^{-10}} = 1.7 \times 10^{-5} \text{ M} = 2 \times 10^{-5} \text{ M}$$

Since 2×10^{-5} M is greater than 1.0×10^{-7} M, the solution is acidic.

c. $[H_3O^+] = 2 \times 10^{-6}$ M, and is greater than 1.0×10^{-7} M. Therefore, the solution is acidic.

d. $[H_3O^+] = 6 \times 10^{-10}$ M, and is less than 1.0×10^{-7} M. Therefore, the solution is basic.

16.49 First, obtain the concentration of H_3O^+ ions using the expression for K_w.

$$[H_3O^+] = \frac{K_w}{[OH^-]} = \frac{1.0 \times 10^{-14}}{1.5 \times 10^{-7}} = 6.67 \times 10^{-8} \text{ M} = 6.7 \times 10^{-8} \text{ M}$$

Since 6.7×10^{-8} M is less than 1.0×10^{-7} M, the solution is basic.

16.50 First, obtain the concentration of H_3O^+ ions using the expression for K_w.

$$[H_3O^+] = \frac{K_w}{[OH^-]} = \frac{1.0 \times 10^{-14}}{8.4 \times 10^{-5}} = 1.19 \times 10^{-10} \text{ M} = 1.2 \times 10^{-10} \text{ M}$$

Since 1.2×10^{-10} M is less than 1.0×10^{-7} M, the solution is basic.

16.51 a. 8.00   b. 2.12   c. 11.30   d. 8.20

16.52 a. 4.00   b. 4.64   c. 9.49   d. 10.54

16.53 The pH of the vinegar, $[H_3O^+] = 7.5 \times 10^{-3}$ M, is 2.12. Since 2.12 is less than 7.00, the solution is acidic.

16.54 The pH of the beer, $[H_3O^+] = 5.0 \times 10^{-3}$ M, is 2.30. Since 2.30 is less than 7.00, the solution is acidic.

CHAPTER 16

16.55 First, find the concentration of H_3O^+ ions using the expression for K_w.

$$[H_3O^+] = \frac{K_w}{[OH^-]} = \frac{1.0 \times 10^{-14}}{0.0040} = 2.50 \times 10^{-12} \text{ M} = 2.5 \times 10^{-12} \text{ M}$$

The pH of this solution is 11.60. Since 11.60 is more than 7.00, the solution is basic.

16.56 First, find the concentration of H_3O^+ ions using the expression for K_w.

$$[H_3O^+] = \frac{K_w}{[OH^-]} = \frac{1.0 \times 10^{-14}}{0.050} = 2.00 \times 10^{-13} \text{ M} = 2.0 \times 10^{-13} \text{ M}$$

The pH of this solution is 12.70. Since 12.70 is more than 7.00, the solution is basic.

16.57 a. $[H_3O^+]$ = inverse log (-7.28) = 5.2×10^{-8} M.

b. $[H_3O^+]$ = inverse log (-2.32) = 4.8×10^{-3} M.

c. $[H_3O^+]$ = inverse log (-12.10) = 7.9×10^{-13} M.

d. $[H_3O^+]$ = inverse log (-6.73) = 1.9×10^{-7} M.

16.58 a. $[H_3O^+]$ = inverse log (-3.41) = 3.9×10^{-4} M.

b. $[H_3O^+]$ = inverse log (-1.56) = 2.8×10^{-2} M.

c. $[H_3O^+]$ = inverse log (-9.96) = 1.1×10^{-10} M.

d. $[H_3O^+]$ = inverse log (-13.35) = 4.5×10^{-14} M.

16.59 $[H_3O^+]$ = inverse log (-8.74) = 1.82×10^{-9} = 1.8×10^{-9} M.

The OH^- concentration is obtained from the expression for K_w.

$$[OH^-] = \frac{K_w}{[H_3O^+]} = \frac{1.0 \times 10^{-14}}{1.82 \times 10^{-9}} = 5.49 \times 10^{-6} \text{ M} = 5.5 \times 10^{-6} \text{ M}$$

16.60 $[H_3O^+]$ = inverse log (-4.74) = 1.82×10^{-5} = 1.8×10^{-5} M.

$$[OH^-] = \frac{K_w}{[H_3O^+]} = \frac{1.0 \times 10^{-14}}{1.82 \times 10^{-5}} = 5.49 \times 10^{-10} \text{ M} = 5.5 \times 10^{-10} \text{ M}$$

Solutions to Additional Problems

Note on significant figures: The first time the final answer is written, it is written first with one nonsignificant figure. The least significant digit is also underlined. The final answer is then rounded to the correct number of significant figures but the least significant digit is no longer underlined.

16.61 $[H_3O^+]$ = inverse log (-2.38) = 4.17×10^{-3} M.

Acetic acid reacts with water to form H_3O^+ ions and $C_2H_3O_2^-$ ions according to

$$HC_2H_3O_2(aq) + H_2O(l) \rightleftharpoons H_3O^+(aq) + C_2H_3O_2^-(aq)$$

Since equal numbers of H_3O^+ ions and $C_2H_3O_2^-$ ions are formed

$$[C_2H_3O_2^-] = [H_3O^+] = 4.\underline{1}7 \times 10^{-3} = 4.2 \times 10^{-3} \text{ M}$$

16.62 $[H_3O^+]$ = inverse log (-11.62) = 2.40×10^{-12} M.

We can find the concentration of OH^- ions using the expression for K_w.

$$[OH^-] = \frac{K_w}{[H_3O^+]} = \frac{1.0 \times 10^{-14}}{2.40 \times 10^{-12}} = 4.17 \times 10^{-3} \text{ M}$$

Ammonia (NH_3) reacts with water according to

$$NH_3(aq) + H_2O(l) \rightleftharpoons NH_4^+(aq) + OH^-(aq)$$

Since equal numbers of NH_4^+ ions and OH^- ions are formed

$$[NH_4^+] = [OH^-] = 4.\underline{1}7 \times 10^{-3} = 4.2 \times 10^{-3} \text{ M}$$

16.63 The concentration of HCl is

$$[HCl] = \frac{\text{mol HCl}}{\text{L solution}} = \frac{1.0 \text{ mol}}{10 \text{ L}} = 0.\underline{1}0 \text{ M} = 0.1 \text{ M}$$

Therefore, since HCl is a strong acid, the hydronium ion concentration is 0.1 M. The pH of the solution is

$$pH = -\log(0.1) = 1.0$$

16.64 First, calculate the concentration of NaOH.

$$[NaOH] = \frac{mol\ NaOH}{L\ solution} = \frac{1.0\ mol}{10\ L} = 0.10\ M$$

Therefore, since NaOH is a strong base, the hydroxide concentration is 0.10 M. Using K_w gives the hydronium ion concentration.

$$[H_3O^+] = \frac{K_w}{[OH^-]} = \frac{1.0 \times 10^{-14}}{0.10} = 1.0 \times 10^{-13}\ M = 1 \times 10^{-13}\ M$$

The pH of the solution is

$$pH = -\log(1 \times 10^{-13}) = 13.0$$

17. OXIDATION-REDUCTION REACTIONS

Solutions to Exercises

17.1 The charge on the nickel ion increases ($Ni^{2+} \longrightarrow Ni^{3+}$) during the reaction. Thus, Ni^{2+} is oxidized and is the reducing agent. The charge on the cadmium ion decreases ($Cd^{2+} \longrightarrow Cd$) during the reaction. Therefore, Cd^{2+} is reduced and is the oxidizing agent.

17.2 Magnesium is above silver in the activity series. Therefore, a reaction between Mg and Ag^+ will occur.

$$Mg(s) + 2\ Ag^+(aq) \longrightarrow Mg^{2+}(aq) + 2\ Ag(s)$$

17.3 Na_2CrO_4: First, we can get the oxidation number for sodium and oxygen from rules 7 and 6 respectively. Thus,

$$\underset{\text{from rule 7}}{\overset{+1}{Na_2}}\underset{}{Cr}\underset{\text{from rule 6}}{\overset{-2}{O_4}}$$

Second, the oxidation number of chromium can be obtained from rule 8 (the sum of the oxidation numbers for the atoms in a compound is zero). In this case, we have

$$2 \times (+1) + \text{oxidation number for chromium} + 4 \times (-2) = 0$$
$$2 + \text{oxidation number for chromium} - 8 = 0$$
$$\text{oxidation number for chromium} = +6$$

In summary,

$$\overset{+1}{Na_2}\overset{+6}{Cr}\overset{-2}{O_4}$$

$Na_2Cr_2O_7$: As in the above example, the oxidation numbers of sodium and oxygen are +1 (rule 7) and -2 (rule 6) respectively. By rule 8 (the sum of the oxidation numbers of the atoms in a compound is zero) we get

$$2 \times (+1) + 2 \times \text{oxidation number of chromium} + 7 \times (-2) = 0$$
$$2 + 2 \times \text{oxidation number of chromium} - 14 = 0$$
$$2 \times \text{oxidation number of chromium} = 12$$
$$\text{oxidation number of chromium} = +6$$

In summary

$$\underset{+1}{Na_2}\underset{+6}{Cr_2}\underset{-2}{O_7}$$

17.4 $S_2O_3^{2-}$: The oxidation number of oxygen is -2 (rule 6). We can obtain the oxidation number of sulfur from rule 8 (the sum of the oxidation numbers for the atoms in a polyatomic ion will equal the charge on the ion). Thus,

$$2 \times \text{oxidation number for S} + 3 \times (-2) = -2$$
$$2 \times \text{oxidation number for S} - 6 = -2$$
$$2 \times \text{oxidation number for S} = +4$$
$$\text{oxidation number for S} = +2$$

17.5 First, you need to assign oxidation numbers to each substance.

$$\underset{+1\ -2}{Na_2O\,(s)} + \underset{+1\ -2}{H_2O\,(l)} \longrightarrow \underset{+1\ -2\ +1}{2\,NaOH\,(aq)}$$

Since there is no change in the oxidation number of any atom (Na is +1, H is +1, and O is -2), this is not an oxidation-reduction reaction.

17.6 Follow the seven steps as outlined in the text.

1. Assign oxidation numbers.

$$\underset{+1\ +5\ -2}{HNO_3} + \underset{+1\ -2}{H_2S} \longrightarrow \underset{+2\ -2}{NO} + \underset{0}{S} + \underset{+1\ -2}{H_2O}$$

OXIDATION-REDUCTION REACTIONS 247

2. Pick atoms with an increase in oxidation number.

$$H_2\overset{-2}{S} \longrightarrow \overset{0}{S}$$

The sulfur atoms are balanced already.

3. Pick atoms with a decrease in oxidation number.

$$H\overset{+5}{N}O_3 \longrightarrow \overset{+2}{N}O$$

The nitrogen atoms are already balanced.

4. Calculate the change in the oxidation numbers.

$$\overset{+5}{HNO_3} + H_2\overset{-2}{S} \longrightarrow \overset{+2}{NO} + \overset{0}{S} + H_2O$$

$1 \times (+2) = +2$
$1 \times (-3) = -3$

5. Make the absolute values of the changes in the oxidation numbers equal.

$$2\,\overset{+5}{HNO_3} + 3\,H_2\overset{-2}{S} \longrightarrow 2\,\overset{+2}{NO} + 3\,\overset{0}{S} + H_2O$$

$3 \times (+2) = +6$
$2 \times (-3) = -6$

6. Balance the remainder of the atoms. To balance hydrogen and oxygen, increase the coefficient of water to 4.

$$2\,HNO_3 + 3\,H_2S \longrightarrow 2\,NO + 3\,S + 4\,H_2O$$

7. Check: There are 8 hydrogen atoms, 2 nitrogen atoms, 6 oxygen atoms, and 3 sulfur atoms on each side. Also, the net charge on each side is 0. Therefore, the equation is balanced.

17.7 Use the nine steps as outlined in the text.

1. Assign oxidation numbers

$$\overset{+1\ -2}{H_2S} + \overset{+5\ -2}{NO_3^-} \longrightarrow \overset{0}{S} + \overset{+2\ -2}{NO}$$

2. Split the skeletal equation into a half-reaction for oxidation and a half-reaction for reduction.

$H_2S \longrightarrow S$ (Oxidation)

$NO_3^- \longrightarrow NO$ (Reduction)

3. Balance all atoms except hydrogen and oxygen. The equations are already balanced.

4. Balance oxygen by adding water.

$H_2S \longrightarrow S$

$NO_3^- \longrightarrow NO + 2 H_2O$

5. Balance hydrogen by adding H^+.

$H_2S \longrightarrow S + 2 H^+$

$NO_3^- + 4 H^+ \longrightarrow NO + 2 H_2O$

6. Balance charge by adding electrons.

$H_2S \longrightarrow S + 2 H^+ + 2 e^-$

$NO_3^- + 4 H^+ + 3 e^- \longrightarrow NO + 2 H_2O$

7. Multiply the half-reactions by factors that will lead to the same number of electrons in each half-reaction.

$3 \times [H_2S \longrightarrow S + 2 H^+ + 2 e^-] = 3 H_2S \longrightarrow 3 S + 6 H^+ + 6 e^-$

$2 \times [NO_3^- + 4 H^+ + 3 e^- \longrightarrow NO + 2 H_2O] =$
$\qquad\qquad\qquad\qquad 2 NO_3^- + 8 H^+ + 6 e^- \longrightarrow 2 NO + 4 H_2O$

8. Add the two half-reactions. Cancel equal amounts of electrons and hydrogen ions on each side.

$$3\ H_2S\ +\ 2\ NO_3^-\ +\ \cancel{2}\overset{2}{8}\ H^+\ +\ \cancel{6\ e^-}\ \longrightarrow\ 3\ S\ +\ 2\ NO\ +\ 4\ H_2O\ +\ \cancel{6\ H^+}\ +\ \cancel{6\ e^-}$$

The result is

$$3\ H_2S\ +\ 2\ NO_3^-\ +\ 2\ H^+\ \longrightarrow\ 3\ S\ +\ 2\ NO\ +\ 4\ H_2O$$

9. Check: There are 8 hydrogen atoms, 3 sulfur atoms, 2 nitrogen atoms, and 6 oxygen atoms on each side. Also, the net charge on each side is zero. Therefore, the equation is balanced. Adding phase labels gives

$$3\ H_2S(aq)\ +\ 2\ NO_3^-(aq)\ +\ 2\ H^+(aq)\ \longrightarrow\ 3\ S(s)\ +\ 2\ NO(g)\ +\ 4\ H_2O(l)$$

17.8 Follow the nine steps outlined in the text.

1. Assign oxidation numbers.

$$MnO_4^-\ +\ I^-\ \longrightarrow\ MnO_2\ +\ I_2$$

Mn: +7, O: −2, I: −1, Mn: +4, O: −2, I: 0

2. Split into half-reactions.

$$I^-\ \longrightarrow\ I_2\quad \text{(Oxidation)}$$

$$MnO_4^-\ \longrightarrow\ MnO_2\quad \text{(Reduction)}$$

3. Balance all atoms except hydrogen and oxygen.

$$2\ I^-\ \longrightarrow\ I_2$$

$$MnO_4^-\ \longrightarrow\ MnO_2$$

4. Balance oxygen by adding water.

$$2\ I^-\ \longrightarrow\ I_2$$

$$MnO_4^-\ \longrightarrow\ MnO_2\ +\ 2\ H_2O$$

5. Balance hydrogen by adding H⁺.

$$2\,I^- \longrightarrow I_2$$

$$MnO_4^- + 4\,H^+ \longrightarrow MnO_2 + 2\,H_2O$$

Since the reaction occurs in basic solution, change the half-reactions from acidic to basic conditions by adding 4 OH⁻ to each side of the reduction half-reaction. Combine the 4 H⁺ and 4 OH⁻ to make 4 H₂O on the left side of the reduction half-reaction. Remove two H₂O from each side to give

$$MnO_4^- + 2\,H_2O \longrightarrow MnO_2 + 4\,OH^-$$

6. Balance the charge by adding electrons.

$$2\,I^- \longrightarrow I_2 + 2\,e^-$$

$$MnO_4^- + 2\,H_2O + 3\,e^- \longrightarrow MnO_2 + 4\,OH^-$$

7. Multiply each half-reaction by a factor that will lead to the same number of electrons in each half-reaction.

$$3 \times [2\,I^- \longrightarrow I_2 + 2\,e^-] = 6\,I^- \longrightarrow 3\,I_2 + 6\,e^-$$

$$2 \times [MnO_4^- + 2\,H_2O + 3\,e^- \longrightarrow MnO_2 + 4\,OH^-] =$$

$$2\,MnO_4^- + 4\,H_2O + 6\,e^- \longrightarrow 2\,MnO_2 + 8\,OH^-$$

8. Add the half-reactions. Cancel 6 electrons from each side.

$$2\,MnO_4^- + 6\,I^- + 4\,H_2O + \cancel{6\,e^-} \longrightarrow 2\,MnO_2 + 3\,I_2 + 8\,OH^- + \cancel{6\,e^-}$$

The result is

$$2\,MnO_4^- + 6\,I^- + 4\,H_2O \longrightarrow 2\,MnO_2 + 3\,I_2 + 8\,OH^-$$

9. Check: There are 2 manganese atoms, 12 oxygen atoms, 6 iodine atoms, and 8 hydrogen atoms on each side. Also, the net charge on each side is -8. Therefore, the equation is balanced. Adding phase labels gives

$$2\ MnO_4^-(aq) + 6\ I^-(aq) + 4\ H_2O(l) \longrightarrow 2\ MnO_2(s) + 3\ I_2(aq) + 8\ OH^-(aq)$$

17.9 Since lead is above copper in the activity series, electrons will flow from lead to copper. Thus, Pb will be oxidized and Cu^{2+} will be reduced. This leads to the half-cell reactions

$$Pb(s) \longrightarrow Pb^{2+}(aq) + 2\ e^- \qquad \text{(oxidation)}$$

$$Cu^{2+} + 2\ e^- \longrightarrow Cu(s) \qquad \text{(reduction)}$$

Since the lead-lead(II) electrode is where oxidation occurs, it is the anode. The copper-copper(II) electrode is where reduction occurs, and is the cathode. Electrons will flow from the lead half-cell through the wire to the copper half-cell. Combining half-reactions gives the overall chemical reaction.

$$Pb(s) + Cu^{2+} \longrightarrow Pb^{2+} + Cu(s)$$

■ Answers to Questions to Test Your Reading

17.1 An <u>oxidation-reduction</u> reaction is a reaction in which electrons are transferred from one substance to another. <u>Oxidation</u> is the loss of electrons while <u>reduction</u> is the gain of electrons. Oxidation cannot occur without reduction. Electrons can only be transferred from one species to another. Thus, one substance must lose electrons and the other gains electrons.

17.2 An <u>oxidizing agent</u> is the participant in an oxidation-reduction reaction that causes the oxidation of another substance. It is reduced during the process. A <u>reducing agent</u> is the participant in an oxidation-reduction reaction that causes the reduction of another substance. It is oxidized in the process.

17.3 The <u>activity series</u> is an arrangement of the elements in order of their ease of losing electrons during reactions in aqueous solutions.

17.4 Since lead is above copper in the table, lead metal will react with Cu^{2+} ions. A free element (in this case Pb) in the table will react with a monatomic ion from another element (in this case Cu^{2+}) in the table if the free element is above the other element in the activity series. Thus

$$Pb(s) + Cu^{2+}(aq) \longrightarrow Pb^{2+}(aq) + Cu(s)$$

17.5 Since tin is above hydrogen in the activity series, tin will dissolve in acid.

$$Sn(s) + 2 H^+(aq) \longrightarrow Sn^{2+}(aq) + H_2(g)$$

17.6 Powdered aluminum would be a reducing agent, since it will lose electrons.

$$Al \longrightarrow Al^{3+} + 3 e^-$$

Since aluminum is being oxidized, it is a reducing agent.

17.7 Since cadmium is above copper in the activity series, the half-reactions that would occur between cadmium and copper are

$$Cd \longrightarrow Cd^{2+} + 2 e^-$$

$$Cu^{2+} + 2 e^- \longrightarrow Cu$$

Thus, copper is not a better reducing agent than cadmium. The reaction is

$$Cd + Cu^{2+} \longrightarrow Cd^{2+} + Cu$$

17.8 An <u>oxidation number</u> is either
i. the charge on an atom or monatomic ion, or
ii. the charge on an atom in a substance if the pairs of electrons in each bond belonged to the more electronegative atom.

This concept is useful as a bookkeeping device to keep track of electrons in oxidation-reduction reactions. It is also helpful in identifying the oxidizing agent and the reducing agent. It is also sometimes used in the naming of compounds.

17.9 The oxidation number of a Cr atom in the element is zero. So are the oxidation numbers of an O atom in O_2, a P atom in P_4, and an S atom in S_8 as these are all elements.

17.10 In $MgCl_2$, the oxidation number of Mg is +2, because it is a Group IIA element. The oxidation number of each Cl atom is -1 because it is a Group VIIA element. The sum of the oxidation numbers in the compound is zero. In HI, hydrogen (Group IA) has an oxidation number of +1 and iodine (Group VIIA) has an oxidation number of -1. The sum of the oxidation numbers in the compound is zero.

17.11 In H_2O, hydrogen has an oxidation number of +1 and oxygen has an oxidation number of -2. The sum of the oxidation numbers in the compound is zero. For hydrogen peroxide, H_2O_2, hydrogen has an oxidation number of +1 and oxygen has an oxidation number of -1. The sum of the oxidation numbers in the compound is zero.

17.12 During oxidation, the oxidation number of an atom increases. During reduction, it decreases.

17.13 The reaction of hydrochloric acid with sodium hydroxide is not an oxidation-reduction reaction. It is a double displacement reaction. No oxidation numbers are changing in this reaction and no electrons are being transferred.

17.14 Coal (carbon) burning in oxygen is an oxidation-reduction reaction. The reaction can be written as

$$C(s) + O_2(g) \longrightarrow CO_2(g)$$

The oxidation numbers of carbon and oxygen in the reactants are both zero (rule 1). In carbon dioxide, oxygen has an oxidation number of -2 and carbon is +4. Note that the oxidation number for carbon changes from 0 to +4, so it is being oxidized by oxygen. The oxidation number of oxygen changes from 0 to -2, so it is being reduced by the carbon. Therefore, oxygen is the oxidizing agent and carbon is the reducing agent.

17.15 <u>Half-reactions</u> are the reactions that result when an oxidation-reduction reaction has been separated into two equations, one involving the species that is oxidized and one involving the species that is reduced. For the reaction

$$Zn(s) + Cu^{2+}(aq) \longrightarrow Zn^{2+}(aq) + Cu(s)$$

The oxidation half-reaction is

$$Zn(s) \longrightarrow Zn^{2+}(aq) + 2\,e^-$$

The reduction half-reaction is

$$Cu^{2+}(aq) + 2\,e^- \longrightarrow Cu(s)$$

Note that half-reactions involve electrons where complete reactions do not.

17.16 The two methods that are used to balance oxidation-reduction reactions are the oxidation number method and the half-reaction method.

17.17 An <u>electrochemical cell</u> is an apparatus that either generates or uses an electric current. A <u>voltaic cell</u> (also called a <u>galvanic cell</u>) is a cell in which a spontaneous reaction generates an electric current. An <u>electrochemical cell</u> is a cell that requires an electric current from some other source to drive an otherwise nonspontaneous reaction.

17.18 A <u>spontaneous process</u> is a physical or chemical change that occurs by itself. An example from everyday life would be the spoiling of food when it is not refrigerated. Food spoils spontaneously. We put it in the refrigerator to try and slow down the process.

17.19 A <u>half-cell</u> is that portion of an electrochemical cell in which a half-reaction takes place. The half-cells in the Daniell cell are

$$Zn(s) \longrightarrow Zn^{2+}(aq) + 2\ e^- \qquad \text{(oxidation)}$$

$$Cu^{2+}(aq) + 2\ e^- \longrightarrow Cu(s) \qquad \text{(reduction)}$$

17.20 The electrode at which oxidation occurs is called the <u>anode</u>. The <u>cathode</u> is the electrode at which reduction occurs. In the Daniell cell, the anode is the zinc-zinc (II) ion half-cell and the cathode is the copper-copper (II) ion half-cell.

17.21 A <u>battery</u> is a series of voltaic cells that are wired so that the anode of one is connected to the cathode of another. Since a Daniell cell generates about 1 volt, a battery consisting of six Daniell cells will have a voltage of 6 x (1V) = 6 V.

17.22 The substance that is oxidized in a lead storage battery is Pb(s). The substance being reduced is PbO_2(s). When this battery is dead, the chemical reaction has reached equilibrium. When it is recharged, an external circuit drives the reaction in the reverse direction, where it is ready to be spontaneous again.

17.23 The reaction that would produce H_2(g) would be

$$2\ H_2O(l) \longrightarrow 2\ H_2(g) + O_2(g)$$

17.24 In a nickel-cadmium battery, the cadmium is oxidized and the NiO(OH) is reduced.

17.25 If you fasten a wire across the terminals of a nickel-cadmium battery, the reaction

$$Cd + 2\ NiO(OH) + 2\ H_2O \longrightarrow Cd(OH)_2 + 2\ Ni(OH)_2$$

will proceed from left to right until the voltage of the cell drops to zero. Also, as a result of the reaction, heat will be generated.

17.26 In a mercury battery, the zinc is oxidized and the mercury is reduced.

■ Solutions to Practice Problems

17.27a.
$$\underset{\text{oxidation}}{\underbrace{Sn^{2+}(aq) + \underset{\text{reduction}}{\underbrace{2\ Ce^{4+}(aq) \longrightarrow Sn^{4+}(aq)}} + 2\ Ce^{3+}(aq)}}$$

Thus, Ce^{4+} is the oxidizing agent and Sn^{2+} is the reducing agent.

b. $Zn(s) + 2H^+(aq) \longrightarrow Zn^{2+}(aq) + H_2(g)$

(reduction: $H^+ \to H_2$; oxidation: $Zn \to Zn^{2+}$)

Thus, H^+ is the oxidizing agent and Zn is the reducing agent.

c. $Pb^{2+}(aq) + Cd(s) \longrightarrow Pb(s) + Cd^{2+}$

(reduction: $Pb^{2+} \to Pb$; oxidation: $Cd \to Cd^{2+}$)

Thus, Pb^{2+} is the oxidizing agent and Cd is the reducing agent.

d. $2Al(s) + 3Cl_2(g) \longrightarrow 2Al^{3+}(aq) + 6Cl^-(aq)$

(reduction: $Cl_2 \to Cl^-$; oxidation: $Al \to Al^{3+}$)

Thus, Cl_2 is the oxidizing agent and Al is the reducing agent.

17.28a. $Cr^{2+}(aq) + Fe^{3+}(aq) \longrightarrow Cr^{3+}(aq) + Fe^{2+}(aq)$

(reduction: $Fe^{3+} \to Fe^{2+}$; oxidation: $Cr^{2+} \to Cr^{3+}$)

Thus, Fe^{3+} is the oxidizing agent and Cr^{2+} is the reducing agent.

b. $I_2(aq) + Fe(s) \longrightarrow 2I^-(aq) + Fe^{2+}$

(reduction: $I_2 \to I^-$; oxidation: $Fe \to Fe^{2+}$)

Thus, I_2 is the oxidizing agent and Fe is the reducing agent.

c.
$$2\text{ H}^+(aq) + \text{Mg}(s) \xrightarrow{\text{reduction / oxidation}} \text{H}_2(g) + \text{Mg}^{2+}(aq)$$

(reduction: H⁺ → H₂; oxidation: Mg → Mg²⁺)

Thus, H⁺ is the oxidizing agent and Mg is the reducing agent.

d.
$$\text{Pb}(s) + \text{Hg}^{2+}(aq) \xrightarrow{\text{reduction / oxidation}} \text{Pb}^{2+}(aq) + \text{Hg}(l)$$

(reduction: Hg²⁺ → Hg; oxidation: Pb → Pb²⁺)

Thus, Hg²⁺ is the oxidizing agent and Pb is the reducing agent.

17.29 a. Lead is below magnesium in the activity series. Therefore, a reaction between Pb(s) and Mg²⁺(aq) will not occur.

b. Zinc is above silver in the activity series. Therefore, a reaction between Zn(s) and Ag⁺(aq) will occur.

c. Chromium is above hydrogen in the activity series. Therefore, a reaction between Cr(s) and H⁺(aq) will occur.

d. Gold is below hydrogen in the activity series. Therefore, a reaction between Au(s) and H⁺(aq) will not occur.

17.30 a. Tin is below zinc in the activity series. Therefore, a reaction between Sn(s) and Zn²⁺(aq) will not occur.

b. Mercury is below hydrogen in the activity series. Therefore, a reaction between Hg(l) and H⁺(aq) will not occur.

c. Hydrogen is below nickel in the activity series. Therefore, a reaction between H₂(g) and Ni²⁺(aq) will not occur.

d. Aluminum is above silver in the activity series. Therefore, a reaction between Al(s) and Ag⁺(aq) will occur.

17.31 a. HBr   b. Na₂O   c. CH₄
  H: +1, Br: −1   Na: +1, O: −2   C: −4, H: +1

d. C$_2$H$_6$ e. O$_2$ f. O$_3$
   −3 +1   0   0

17.32 a. NH$_3$ b. H$_2$O   c. Al$_2$O$_3$
   −3 +1   +1 −2   +3 −2

d. Al(OH)$_3$ e. CrO   f. CrO$_3$
   +3 −2 +1   +2 −2   +6 −2

17.33 a. NaClO$_4$ b. NaClO$_3$ c. NaClO$_2$
   +1 +7 −2   +1 +5 −2   +1 +3 −2

d. NaClO   e. NaCl   f. Cl$_2$
   +1 +1 −2   +1 −1   0

17.34 a. BaCrO$_4$ b. Sr(MnO$_4$)$_2$ c. Al(NO$_3$)$_3$
   +2 +6 −2   +2 +7 −2   +3 +5 −2

d. H$_3$PO$_4$ e. H$_2$SO$_4$ f. P$_4$
   +1 +5 −2   +1 +6 −2   0

17.35 a. Br$^-$ b. N^{3-} c. NH$_2^-$
   −1   −3   −3 +1

d. OH$^-$ e. MnO$_4^-$ f. HS$^-$
   −2 +1   +7 −2   +1 −2

17.36 a. S^{2-} : -2
b. SO_3^{2-} : $+4, -2$
c. HSO_3^- : $+1, +4, -2$

d. HSO_4^- : $+1, +6, -2$
e. CO_3^{2-} : $+4, -2$
f. HCO_3^- : $+1, +4, -2$

17.37 a. $H_2PO_4^-$: $+1, +5, -2$
b. HPO_4^{2-} : $+1, +5, -2$
c. PO_4^{3-} : $+5, -2$

d. P^{3-} : -3
e. NO_3^- : $+5, -2$
f. NO_2^- : $+3, -2$

17.38 a. CrO_4^{2-} : $+6, -2$
b. $HCrO_4^-$: $+1, +6, -2$
c. $Cr_2O_7^{2-}$: $+6, -2$

d. VO^{2+} : $+4, -2$
e. VO_2^+ : $+5, -2$
f. IO_3^- : $+5, -2$

17.39a. Assign oxidation numbers:

$$N_2(g) + O_2(g) \longrightarrow 2\,NO(g)$$
$$0 \qquad\quad 0 \qquad\qquad\quad +2\ \ -2$$

Since the oxidation number of nitrogen increases, it is being oxidized. Oxygen is being reduced. Therefore, N_2 is the reducing agent and O_2 is the oxidizing agent.

b. Assign oxidation numbers:

$$P_4(s) + 5\ O_2(g) \longrightarrow P_4O_{10}(s)$$
$$0 \qquad0 \qquad\qquad +5\ \ -2$$

Since the oxidation number of phosphorus increases, it is being oxidized. Oxygen is being reduced. Therefore, P_4 is the reducing agent and O_2 is the oxidizing agent.

c. Assign oxidation numbers:

$$P_4O_{10}(s) + 6\ H_2O(l) \longrightarrow 4\ H_3PO_4(aq)$$
$$+5\ \ -2 \qquad +1\ \ -2 \qquad\qquad +1\ +5\ \ -2$$

Since the oxidation numbers for phosphorus, hydrogen, and oxygen do not change during the reaction, this equation does not represent an oxidation-reduction reaction.

d. Assign oxidation numbers:

$$2\ CO(g) + O_2(g) \longrightarrow 2\ CO_2(g)$$
$$+2\ \ -2 \qquad 0 \qquad\qquad +4\ \ -2$$

Since the oxidation number of carbon increases, it is being oxidized. Oxygen is being reduced. Therefore, CO is the reducing agent and O_2 is the oxidizing agent.

17.40a. Assign oxidation numbers:

$$N_2(g) + 3\ H_2(g) \longrightarrow 2\ NH_3(g)$$
$$0 \qquad0 \qquad\qquad -3\ \ +1$$

Since the oxidation number of hydrogen increases, it is being oxidized. Nitrogen is being reduced. Thus, N_2 is the oxidizing agent and H_2 is the reducing agent.

b. Assign oxidation numbers:

$$K_2O(s) + H_2O(l) \longrightarrow 2\ KOH(aq)$$
$$+1\ -2 \qquad +1\ -2 \qquad\qquad +1\ -2\ +1$$

Since the oxidation numbers of potassium, oxygen, and hydrogen do not change during the reaction, this equation does not represent an oxidation-reduction reaction.

260 ■ CHAPTER 17

c. Assign oxidation numbers:

$$2\ Na(s) + 2\ H_2O(l) \longrightarrow 2\ NaOH(aq) + H_2(g)$$

Na: 0, H: +1, O: -2, Na: +1, O: -2, H: +1, H: 0

Since the oxidation number of sodium increases, it is being oxidized. Water is being reduced. Therefore, Na is the reducing agent and H_2O is the oxidizing agent.

d. Assign oxidation numbers:

$$C_2H_4(g) + H_2(g) \longrightarrow C_2H_6(g)$$

C: -2, H: +1, H: 0, C: -3, H: +1

Since the oxidation number of hydrogen is increasing, it is being oxidized. Carbon is being reduced. Therefore, H_2 is the reducing agent and C_2H_4 is the oxidizing agent.

17.41 a. $4\ Li(s) + O_2(g) \longrightarrow 2\ Li_2O(s)$

O_2 is the oxidizing agent and Li is the reducing agent.

b. $4\ Al(s) + 3\ O_2(g) \longrightarrow 2\ Al_2O_3(s)$

O_2 is the oxidizing agent and Al is the reducing agent.

c. $2\ Cr(s) + 6\ HCl(aq) \longrightarrow 2\ CrCl_3(aq) + 3\ H_2(g)$

HCl is the oxidizing agent and Cr is the reducing agent.

d. $CH_4(g) + 2\ O_2(g) \longrightarrow CO_2(g) + 2\ H_2O(l)$

O_2 is the oxidizing agent and CH_4 is the reducing agent.

17.42 a. $2\ Na(s) + O_2(g) \longrightarrow Na_2O_2(s)$

O_2 is the oxidizing agent and Na is the reducing agent.

OXIDATION-REDUCTION REACTIONS ■ 261

b. 8 Fe(s) + S$_8$(l) ⟶ 8 FeS(s)

S$_8$ is the oxidizing agent and Fe is the reducing agent.

c. 2 B(s) + 3 H$_2$(g) ⟶ B$_2$H$_6$(g)

H$_2$ is the oxidizing agent and B is the reducing agent.

d. C$_2$H$_4$(g) + 3 O$_2$(g) ⟶ 2 CO$_2$(g) + 2 H$_2$O(l)

O$_2$ is the oxidizing agent and C$_2$H$_4$ is the reducing agent.

17.43a. Br$_2$(aq) + SO$_2$(aq) + 2 H$_2$O(l) ⟶ 2 HBr(aq) + H$_2$SO$_4$(aq)

Br$_2$ is the oxidizing agent and SO$_2$ is the reducing agent.

b. 6 HI(aq) + 2 HNO$_3$(aq) ⟶ 3 I$_2$(aq) + 2 NO(g) + 4 H$_2$O(l)

HNO$_3$ is the oxidizing agent and HI is the reducing agent.

c. MnO$_2$(s) + 4 HBr(aq) ⟶ MnBr$_2$(aq) + Br$_2$(aq) + 2 H$_2$O(l)

MnO$_2$ is the oxidizing agent and HBr is the reducing agent.

d. 3 Ag(s) + 4 HNO$_3$(aq) ⟶ 3 AgNO$_3$(aq) + NO(g) + 2 H$_2$O(l)

HNO$_3$ is the oxidizing agent and Ag is the reducing agent.

17.44a. 24 H$_2$S(aq) + 16 HNO$_3$(aq) ⟶ 3 S$_8$(s) + 16 NO(g) + 32 H$_2$O(l)

HNO$_3$ is the oxidizing agent and H$_2$S is the reducing agent.

b. H$_2$SO$_4$(l) + 2 HBr(g) ⟶ SO$_2$(g) + Br$_2$(g) + 2 H$_2$O(g)

H$_2$SO$_4$ is the oxidizing agent and HBr is the reducing agent.

c. $16\ HCl(aq)\ +\ 2\ KMnO_4(aq) \longrightarrow$

$5\ Cl_2(g)\ +\ 2\ MnCl_2(aq)\ +\ 2\ KCl(aq)\ +\ 8\ H_2O(l)$

$KMnO_4$ is the oxidizing agent and HCl is the reducing agent.

d. $H_2S(aq)\ +\ 4\ I_2(aq)\ +\ 4\ H_2O(l) \longrightarrow H_2SO_4(aq)\ +\ 8\ HI(aq)$

I_2 is the oxidizing agent and H_2S is the reducing agent.

17.45a. $Cr_2O_7^{2-}(aq)\ +\ 6\ Fe^{2+}(aq)\ +\ 14\ H^+(aq) \longrightarrow$

$2\ Cr^{3+}(aq)\ +\ 6\ Fe^{3+}(aq)\ +\ 7\ H_2O(l)$

$Cr_2O_7^{2-}$ is the oxidizing agent and Fe^{2+} is the reducing agent.

b. $2\ VO_2^+(aq)\ +\ Zn(s)\ +\ 4\ H^+(aq) \longrightarrow 2\ VO^{2+}(aq)\ +\ Zn^{2+}(aq)\ +\ 2\ H_2O(l)$

VO_2^+ is the oxidizing agent and Zn is the reducing agent.

c. $VO_2^+(aq)\ +\ Zn(s)\ +\ 4\ H^+(aq) \longrightarrow V^{3+}(aq)\ +\ Zn^{2+}(aq)\ +\ 2\ H_2O(l)$

VO_2^+ is the oxidizing agent and Zn is the reducing agent.

d. $2\ VO_2^+(aq)\ +\ 3\ Zn(s)\ +\ 8\ H^+(aq) \longrightarrow 2\ V^{2+}(aq)\ +\ 3\ Zn^{2+}(aq)\ +\ 4\ H_2O(l)$

VO_2^+ is the oxidizing agent and Zn is the reducing agent.

17.46a. $2\ MnO_4^-(aq)\ +\ 5\ Sn^{2+}(aq)\ +\ 16\ H^+(aq) \longrightarrow$

$2\ Mn^{2+}(aq)\ +\ 5\ Sn^{4+}(aq)\ +\ 8\ H_2O(l)$

MnO_4^- is the oxidizing agent and Sn^{2+} is the reducing agent.

b. $Cl_2(g) + IO_3^-(aq) + H_2O(l) \longrightarrow 2\ Cl^-(aq) + IO_4^-(aq) + 2\ H^+(aq)$

Cl_2 is the oxidizing agent and IO_3^- is the reducing agent.

c. $Cu_2O(s) + Cl_2(g) + 2\ H^+(aq) \longrightarrow 2\ Cu^{2+}(aq) + 2\ Cl^-(aq) + H_2O(l)$

Cl_2 is the oxidizing agent and Cu_2O is the reducing agent.

d. $I_2(aq) + 5\ Cl_2(g) + 6\ H_2O(l) \longrightarrow 2\ HIO_3(aq) + 10\ Cl^-(aq) + 10\ H^+(aq)$

Cl_2 is the oxidizing agent and I_2 is the reducing agent.

17.47a. $48\ NO_3^-(aq) + S_8(s) + 32\ H^+(aq) \longrightarrow 48\ NO_2(g) + 8\ SO_4^{2-}(aq) + 16\ H_2O(l)$

NO_3^- is the oxidizing agent and S_8 is the reducing agent.

b. $2\ MnO_4^-(aq) + 5\ SO_2(g) + 2\ H_2O(l) \longrightarrow 2\ Mn^{2+}(aq) + 5\ SO_4^{2-}(aq) + 4\ H^+(aq)$

MnO_4^- is the oxidizing agent and SO_2 is the reducing agent.

c. $Cr_2O_7^{2-}(aq) + 3\ HNO_2(aq) + 5\ H^+(aq) \longrightarrow 2\ Cr^{3+}(aq) + 3\ NO_3^-(aq) + 4\ H_2O(l)$

$Cr_2O_7^{2-}$ is the oxidizing agent and HNO_2 is the reducing agent.

d. $2\ MnO_4^-(aq) + 5\ HNO_2(aq) + H^+(aq) \longrightarrow 2\ Mn^{2+}(aq) + 5\ NO_3^-(aq) + 3\ H_2O(l)$

MnO_4^- is the oxidizing agent and HNO_2 is the reducing agent.

17.48a. $3\ MnO_4^-(aq) + 5\ V^{2+}(aq) + 4\ H^+(aq) \longrightarrow 3\ Mn^{2+}(aq) + 5\ VO_2^+(aq) + 2\ H_2O(l)$

MnO_4^- is the oxidizing agent and V^{2+} is the reducing agent.

b. $2\ MnO_4^-(aq) + 5\ V^{3+}(aq) + 2\ H_2O(l) \longrightarrow 2\ Mn^{2+}(aq) + 5\ VO_2^+(aq) + 4\ H^+(aq)$

MnO_4^- is the oxidizing agent and V^{3+} is the reducing agent.

c. $MnO_4^-(aq) + 5\ VO^{2+}(aq) + H_2O(l) \longrightarrow Mn^{2+}(aq) + 5\ VO_2^+(aq) + 2\ H^+(aq)$

MnO_4^- is the oxidizing agent and VO^{2+} is the reducing agent.

d. $2\ MnO_4^-(aq) + 5\ C_2O_4^{2-}(aq) + 16\ H^+(aq) \longrightarrow$

$2\ Mn^{2+}(aq) + 10\ CO_2(g) + 8\ H_2O(l)$

MnO_4^- is the oxidizing agent and $C_2O_4^{2-}$ is the reducing agent.

17.49a. $8\ Al(s) + 3\ NO_3^-(aq) + 5\ OH^-(aq) + 18\ H_2O(l) \longrightarrow 8\ Al(OH)_4^-(aq) + 3\ NH_3(aq)$

NO_3^- is the oxidizing agent and Al is the reducing agent.

b. $S^{2-}(aq) + 4\ I_2(aq) + 8\ OH^-(aq) \longrightarrow SO_4^{2-}(aq) + 8\ I^-(aq) + 4\ H_2O(l)$

I_2 is the oxidizing agent and S^{2-} is the reducing agent.

c. $Mn(OH)_2(s) + H_2O_2(aq) \longrightarrow MnO_2(s) + 2\ H_2O(l)$

H_2O_2 is the oxidizing agent and $Mn(OH)_2$ is the reducing agent.

d. $P_4(s) + 4\ OH^-(aq) + 2\ H_2O(l) \longrightarrow 2\ PH_3(g) + 2\ HPO_3^{2-}(aq)$

P_4 is both the oxidizing agent and the reducing agent.

17.50a. $Pb(OH)_4^{2-}(aq) + ClO^-(aq) \longrightarrow PbO_2(s) + Cl^-(aq) + 2\ OH^-(aq) + H_2O(l)$

ClO^- is the oxidizing agent and $Pb(OH)_4^{2-}$ is the reducing agent.

b. $IO_3^-(aq) + 5\ I^-(aq) + 3\ H_2O(l) \longrightarrow 3\ I_2(aq) + 6\ OH^-(aq)$

IO_3^- is the oxidizing agent and I^- is the reducing agent.

c. $2\ Cr(OH)_2(s) + ClO^-(aq) + H_2O(l) \longrightarrow 2\ Cr(OH)_3(s) + Cl^-(aq)$

ClO$^-$ is the oxidizing agent and Cr(OH)$_2$ is the reducing agent.

d. $CrO_4^{2-}(aq) + 3\ Fe(OH)_2(s) + 4\ H_2O(l) \longrightarrow Cr(OH)_3(s) + 3\ Fe(OH)_3(s) + 2\ OH^-(aq)$

CrO$_4^{2-}$ is the oxidizing agent and Fe(OH)$_2$ is the reducing agent.

17.51 Since tin is above copper in the activity series, electrons will flow from the tin to the copper ions. Thus, Sn will be oxidized and Cu^{2+} will be reduced. Since oxidation occurs at the tin-tin(II) electrode, it is the anode. The copper-copper(II) electrode is the cathode since reduction occurs there. The overall reaction is

$$Sn(s) + Cu^{2+}(aq) \longrightarrow Sn^{2+}(aq) + Cu(s)$$

17.52 Since zinc is above iron in the activity series, electrons will flow from the zinc to the iron ions. Thus, Zn will be oxidized and Fe^{2+} will be reduced. Since oxidation occurs at the zinc-zinc(II) electrode, it is the anode. The iron-iron(II) electrode is the cathode, since reduction occurs there. The overall reaction is

$$Zn(s) + Fe^{2+}(aq) \longrightarrow Zn^{2+}(aq) + Fe(s)$$

17.53 Chromium is above lead on the activity series. Therefore, chromium will be oxidized and lead will be reduced. The anode and cathode are

$$Pb^{2+} + 2\ e^- \longrightarrow Pb \quad \text{(cathode)}$$

$$Cr \longrightarrow Cr^{2+} + 2\ e^- \quad \text{(anode)}$$

The overall reaction is

$$Cr(s) + Pb^{2+}(aq) \longrightarrow Cr^{2+}(aq) + Pb(s)$$

17.54 Cadmium is above nickel on the activity series. Therefore, cadmium will be oxidized and nickel will be reduced. The anode and cathode are

$$Cd \longrightarrow Cd^{2+} + 2\ e^- \quad \text{(anode)}$$

$$Ni^{2+} + 2\ e^- \longrightarrow Ni \quad \text{(cathode)}$$

The overall reaction is

$$Cd(s) + Ni^{2+}(aq) \longrightarrow Cd^{2+}(aq) + Ni(s)$$

Solutions to Additional Problems

17.55 $2\ Cr(OH)_3(s)\ +\ 3\ H_2O_2(aq)\ +\ 4\ KOH(aq)\ \longrightarrow\ 2\ K_2CrO_4(aq)\ +\ 8\ H_2O(l)$

17.56 $3\ H_2O_2(aq)\ +\ 2\ KMnO_4(aq)\ \longrightarrow\ 2\ MnO_2(s)\ +\ 3\ O_2(g)\ +\ 2\ KOH(aq)\ +\ 2\ H_2O(l)$

17.57 Follow the steps in the text.

<u>Oxidation Number Method</u>

1. Assign oxidation numbers.

$$\underset{+1}{Ag^+}\ +\ \underset{0}{Sn}\ \longrightarrow\ \underset{0}{Ag}\ +\ \underset{+2}{Sn^{2+}}$$

2. Pick atoms with an increase in oxidation number.

$$\underset{0}{Sn}\ \longrightarrow\ \underset{+2}{Sn^{2+}}$$

3. Pick atoms with a decrease in oxidation number.

$$\underset{+1}{Ag^+}\ \longrightarrow\ \underset{0}{Ag}$$

4. Calculate the changes in oxidation number.

$$\underset{+1}{Ag^+}\ +\ \underset{0}{Sn}\ \longrightarrow\ \underset{0}{Ag}\ +\ \underset{+2}{Sn^{2+}}$$

$1 \times (+2) = +2$

$1 \times (-1) = -1$

5. Make the absolute values of the changes in oxidation number equal.

$$\underset{+1}{2\ Ag^+}\ +\ \underset{0}{Sn}\ \longrightarrow\ \underset{0}{2\ Ag}\ +\ \underset{+2}{Sn^{2+}}$$

$1 \times (+2) = +2$

$2 \times (-1) = -2$

OXIDATION-REDUCTION REACTIONS

6. All remaining atoms are balanced.

$$2\,Ag^+ + Sn \longrightarrow 2\,Ag + Sn^{2+}$$

7. Check: There are 2 silver atoms and 1 tin atom on each side of the equation. Also, the net charge on each side is +2. Therefore, the equation is balanced. After adding phase labels, you get

$$2\,Ag^+(aq) + Sn(s) \longrightarrow 2\,Ag(s) + Sn^{2+}(aq)$$

Half-Reaction Method

1. Assign oxidation numbers.

$$\underset{+1}{Ag^+} + \underset{0}{Sn} \longrightarrow \underset{0}{Ag} + \underset{+2}{Sn^{2+}}$$

2. Separate into half-reactions.

$$Sn \longrightarrow Sn^{2+} \qquad \text{(Oxidation)}$$

$$Ag^+ \longrightarrow Ag \qquad \text{(Reduction)}$$

3, 4, and 5. All atoms are already balanced.

6. Balance charge by adding electrons.

$$Ag^+ + e^- \longrightarrow Ag$$

$$Sn \longrightarrow Sn^{2+} + 2\,e^-$$

7. Multiply each half-reaction by a factor that will lead to the same number of electrons in each half-reaction.

$$2 \times [Ag^+ + e^- \longrightarrow Ag] = 2\,Ag^+ + 2\,e^- \longrightarrow 2\,Ag$$

$$1 \times [Sn \longrightarrow Sn^{2+} + 2\,e^-] = Sn \longrightarrow Sn^{2+} + 2\,e^-$$

268 ■ CHAPTER 17

8. Add the half-reactions. Cancel two electrons from each side.

$$2\ Ag^+ + Sn + \cancel{2e^-} \longrightarrow 2\ Ag + Sn^{2+} + \cancel{2e^-}$$

The result is

$$2\ Ag^+ + Sn \longrightarrow 2\ Ag + Sn^{2+}$$

9. Check: There are 2 silver atoms and 1 tin atom on each side of the equation. Also, the net charge on each side is +2. Therefore, the equation is balanced. After adding phase labels, you get

$$2\ Ag^+(aq) + Sn(s) \longrightarrow 2\ Ag(s) + Sn^{2+}(aq)$$

17.58 Oxidation Number Method.

1. Assign oxidation numbers.

$$\underset{+2}{Cu^{2+}} + \underset{-1}{I^-} \longrightarrow \underset{+1\ -1}{CuI} + \underset{0}{I_2}$$

2. Pick atoms with an increase in oxidation number.

$$\underset{-1}{2\ I^-} \longrightarrow \underset{0}{I_2}$$

The iodine atoms have been balanced.

3. Pick atoms with a decrease in oxidation number.

$$\underset{+2}{Cu^{2+}} \longrightarrow \underset{+1}{CuI}$$

4. Calculate the changes in oxidation number.

$$\overset{1 \times (-1) = -1}{\underset{2 \times (+1) = +2}{\underset{+2}{Cu^{2+}} + \underset{-1}{2\ I^-} \longrightarrow \underset{+1}{CuI} + \underset{0}{I_2}}}$$

5. Make the absolute values of the changes in the oxidation numbers equal.

$$\overset{\overset{2\times(-1)=-2}{\longrightarrow}}{2\overset{+2}{Cu^{2+}} + 2\underset{-1}{I^{-}} \longrightarrow 2\overset{+1}{CuI} + \underset{\underset{2\times(+1)=+2}{\longleftarrow}}{\overset{0}{I_2}}}$$

6. Balance iodine atoms.

$$2\,Cu^{2+} + 4\,I^{-} \longrightarrow 2\,CuI + I_2$$

7. Check: There are 2 copper atoms and 4 iodine atoms on each side of the equation. Also, the net charge on each side is 0. Therefore, the equation is balanced. After adding phase labels, you get

$$2\,Cu^{2+}(aq) + 4\,I^{-}(aq) \longrightarrow 2\,CuI(s) + I_2(aq)$$

<u>Half-Reaction Method</u>.

1. Assign oxidation numbers.

$$\underset{+2}{Cu^{2+}} + \underset{-1}{I^{-}} \longrightarrow \underset{+1\;-1}{CuI} + \underset{0}{I_2}$$

2. Break into half-reactions.

$$I^{-} \longrightarrow I_2 \qquad \text{(Oxidation)}$$

$$Cu^{2+} \longrightarrow CuI \qquad \text{(Reduction)}$$

3. Balance iodine atoms.

$$Cu^{2+} + I^{-} \longrightarrow CuI$$

$$2\,I^{-} \longrightarrow I_2$$

4 and 5. Oxygen and hydrogen are already balanced.

6. Balance charge by adding electrons.

$$Cu^{2+} + I^- + e^- \longrightarrow CuI$$

$$2 I^- \longrightarrow I_2 + 2 e^-$$

7. Multiply each half-reaction by a factor that will lead to the same number of electrons in each half-reaction.

$$2 \times [e^- + Cu^{2+} + I^- \longrightarrow CuI] = 2 e^- + 2 Cu^{2+} + 2 I^- \longrightarrow 2 CuI$$

$$1 \times [2 I^- \longrightarrow I_2 + 2 e^-] = 2 I^- \longrightarrow I_2 + 2 e^-$$

8. Add the half-reactions. Cancel two electrons from each side.

$$2 Cu^{2+} + 4 I^- + \cancel{2 e^-} \longrightarrow 2 CuI + I_2 + \cancel{2 e^-}$$

The result is

$$2 Cu^{2+} + 4 I^- \longrightarrow 2 CuI + I_2$$

9. Check: There are 2 copper atoms and 4 iodine atoms on each side of the equation. Also, the net charge on each side is 0. Therefore, the equation is balanced. After adding phase labels, you get

$$2 Cu^{2+}(aq) + 4 I^-(aq) \longrightarrow 2 CuI(s) + I_2(aq)$$

17.59 $4 Fe(OH)_2(s) + O_2(g) + 2 H_2O(l) \longrightarrow 4 Fe(OH)_3(s)$

17.60 $2 Bi^{3+}(aq) + 3 SnO_2^{2-}(aq) + 6 OH^-(aq) \longrightarrow$

$$2 Bi(s) + 3 SnO_3^{2-}(aq) + 3 H_2O(l)$$

17.61 $6 CO_2(g) + 6 H_2O(l) \longrightarrow C_6H_{12}O_6(s) + 6 O_2(g)$

CO_2 is the oxidizing agent and H_2O is the reducing agent.

17.62 2 Al(s) + 6 H$_2$O(l) + 2 OH$^-$(aq) $\longrightarrow$ 2 Al(OH)$_4^-$(aq) + 3 H$_2$(g)

H$_2$O is the oxidizing agent and Al is the reducing agent.

17.63 Follow the steps as outlined in the text.

1. Assign oxidation numbers

    HCHO$_2$ + Ce^{4+} $\longrightarrow$ CO$_2$ + Ce^{3+}
    +1 +2 +1 −2   +4   +4 −2   +3

2. Split into half-reactions.

    HCHO$_2$ $\longrightarrow$ CO$_2$ (Oxidation)

    Ce^{4+} $\longrightarrow$ Ce^{3+} (Reduction)

3 and 4. All atoms, except hydrogen, are already balanced.

5. Balance hydrogen by adding H$^+$.

    HCHO$_2$ $\longrightarrow$ CO$_2$ + 2 H$^+$

    Ce^{4+} $\longrightarrow$ Ce^{3+}

6. Balance charge by adding e$^-$.

    HCHO$_2$ $\longrightarrow$ CO$_2$ + 2 H$^+$ + 2 e$^-$

    Ce^{4+} + e$^-$ $\longrightarrow$ Ce^{3+}

7. Multiply each half-reaction by a factor that will lead to the same number of electrons in each half-reaction.

    1 × [HCHO$_2$ $\longrightarrow$ CO$_2$ + 2 H$^+$ + 2 e$^-$] =

    $\qquad\qquad\qquad\qquad$ HCHO$_2$ $\longrightarrow$ CO$_2$ + 2 H$^+$ + 2 e$^-$

    2 × [Ce^{4+} + e$^-$ $\longrightarrow$ Ce^{3+}] = 2 Ce^{4+} + 2 e$^-$ $\longrightarrow$ 2 Ce^{3+}

8. Add the half-reactions. Cancel two electrons from each side.

$HCHO_2 + 2\ Ce^{4+} + \cancel{2e^-} \longrightarrow CO_2 + 2\ Ce^{3+} + 2\ H^+ + \cancel{2e^-}$

The result is

$HCHO_2 + 2\ Ce^{4+} \longrightarrow CO_2 + 2\ Ce^{3+} + 2\ H^+$

9. Check: There are 2 hydrogen atoms, 1 carbon atom, 2 oxygen atoms, and 2 cerium atoms on each side of the reaction. Also, the net charge on each side is +8. Therefore, the equation is balanced. After adding phase labels, you get

$HCHO_2(aq) + 2\ Ce^{4+}(aq) \longrightarrow CO_2(aq) + 2\ Ce^{3+}(aq) + 2\ H^+(aq)$

17.64 Follow the steps as outlined in the book.

1. Assign oxidation numbers.

$Cl_2 \longrightarrow Cl^- + ClO_3^-$
  0   -1   +5 -2

2. Break into half reactions.

$Cl_2 \longrightarrow ClO_3^-$ (Oxidation)

$Cl_2 \longrightarrow Cl^-$ (Reduction)

3. Balance chlorine atoms.

$Cl_2 \longrightarrow 2\ Cl^-$

$Cl_2 \longrightarrow 2\ ClO_3^-$

4. Balance oxygen by adding water.

$Cl_2 \longrightarrow 2\ Cl^-$

$Cl_2 + 6\ H_2O \longrightarrow 2\ ClO_3^-$

5. Balance hydrogen by adding H^+.

$$Cl_2 \longrightarrow 2\,Cl^-$$

$$Cl_2 + 6\,H_2O \longrightarrow 2\,ClO_3^- + 12\,H^+$$

Since the reaction occurs in basic conditions, add 12 OH^- to each side of the oxidation half-reaction. Combine the 12 OH^- and 12 H^+ to make 12 H_2O. Eliminate 6 H_2O from each side of the half-reaction to give

$$Cl_2 + 12\,OH^- \longrightarrow 2\,ClO_3^- + 6\,H_2O$$

6. Balance charge by adding e^-.

$$Cl_2 + 2\,e^- \longrightarrow 2\,Cl^-$$

$$Cl_2 + 12\,OH^- \longrightarrow 2\,ClO_3^- + 6\,H_2O + 10\,e^-$$

7. Multiply each half-reaction by a factor that will lead to the same number of electrons in each half-reaction.

$$5 \times [Cl_2 + 2\,e^- \longrightarrow 2\,Cl^-] = 5\,Cl_2 + 10\,e^- \longrightarrow 10\,Cl^-$$

$$1 \times [Cl_2 + 12\,OH^- \longrightarrow 2\,ClO_3^- + 6\,H_2O + 10\,e^-] =$$

$$Cl_2 + 12\,OH^- \longrightarrow 2\,ClO_3^- + 6\,H_2O + 10\,e^-$$

8. Add the half-reactions. Cancel 10 e^- from each side.

$$6\,Cl_2 + 12\,OH^- + \cancel{10\,e^-} \longrightarrow 10\,Cl^- + 2\,ClO_3^- + 6\,H_2O + \cancel{10\,e^-}$$

Also, each coefficient in the equation can be divided by two. The result is

$$3\,Cl_2 + 6\,OH^- \longrightarrow 5\,Cl^- + ClO_3^- + 3\,H_2O$$

9. Check: There are 6 chlorine atoms, 6 oxygen atoms, and 6 hydrogen atoms on each side of the reaction. Also, the net charge on each side is -6. Therefore, the equation is balanced. After adding phase labels, you get

$$3\,Cl_2(aq) + 6\,OH^-(aq) \longrightarrow 5\,Cl^-(aq) + ClO_3^-(aq) + 3\,H_2O(l)$$

18. NUCLEAR CHEMISTRY

■ Solutions to Exercises

18.1 The atomic numbers of potassium and calcium are 19 and 20, respectively. Their symbols are $^{40}_{19}K$ and $^{40}_{20}Ca$. The symbol for a beta particle is $^{0}_{-1}e$. The correct equation is

$$^{40}_{19}K \longrightarrow {}^{40}_{20}Ca + {}^{0}_{-1}e$$

18.2 The atomic number of plutonium is 94. The symbol for plutonium-239 is $^{239}_{94}Pu$. The symbol for an alpha particle is $^{4}_{2}He$. If we write $^{A}_{Z}X$ for the unknown nuclide, the nuclear equation is

$$^{239}_{94}Pu \longrightarrow {}^{A}_{Z}X + {}^{4}_{2}He$$

Since A and Z must be conserved, the values of A on both sides of the equation must give the same sum, as must the value of Z. This gives

$$\begin{aligned} A: &\quad 239 = A + 4, \quad \text{or } A = 235 \\ Z: &\quad 94 = Z + 2, \quad \text{or } Z = 92 \end{aligned}$$

Therefore, the product nuclide must be uranium, with atomic number 92. The symbol is $^{235}_{92}U$. The nuclear equation is

$$^{239}_{94}Pu \longrightarrow {}^{235}_{92}U + {}^{4}_{2}He$$

18.3 If we let the original nuclide be represented by $^{A}_{Z}X$ the nuclear equation can be written as

$$^{A}_{Z}X + {}^{4}_{2}He \longrightarrow {}^{30}_{15}P + {}^{1}_{0}n$$

Since A and Z must be conserved, we can write

$$\begin{aligned} A: &\quad A + 4 = 30 + 1, \quad \text{or } A = 27 \\ Z: &\quad Z + 2 = 15 + 0, \quad \text{or } Z = 13 \end{aligned}$$

Aluminum is element 13. Thus the original nucleus must be $^{27}_{13}Al$

18.4 After one half-life, ½ of the original sample will remain. After two half-lives, ½ x ½ = ¼ of the original atoms will remain. Following this pattern, after four half-lives, ½ x ½ x ½ x ½ = 1/16 of the original sample will be left. Therefore, the number of atoms remaining after four half-lives will be

$$1/16 \times 32{,}000{,}000 \text{ atoms} = 2{,}000{,}000 \text{ atoms}$$

18.5 We must first determine how many half-lives have passed. If the original number of disintegrations per minute is 16, then after one half-life it will be reduced to 8 disintegrations per minute, and after two half-lives, 4 disintegrations per minute. Thus, two half-lives must have passed. Since each half-life is 5700 years, the total amount of time that has passed since the animal died is

$$2 \times 5700 \text{ years} = 11{,}400 \text{ years}$$

Thus, the animal must have died in

$$2000 - 11{,}400 = -9400 \text{ AD} \quad \text{or} \quad 9400 \text{ BC}$$

■ Answers to Questions to Test Your Reading

18.1 <u>Radioactivity</u> is the emission of radiation from the nuclei of atoms undergoing spontaneous disintegration.

18.2 The atomic number of nitrogen-14 (and all other forms of nitrogen) is 7. Its mass number is 14. The number of protons in the nucleus is equal to the atomic number, 7. The number of neutrons is 14 - 7 or 7 and the number of nucleons is equal to the mass number, 14. The correct symbol is $^{14}_{7}N$.

18.3 <u>Isotopes</u> are atoms whose nuclei have the same atomic number but different mass number. A <u>nuclide</u> is an nucleus of a specific isotope.

18.4 <u>Radioactive decay</u> is the disintegration of a nucleus with the discharge of radiation and the formation of a new nucleus. The four types of radiation that are emitted during radioactive decay are <u>alpha emission</u>, <u>beta emission</u>, <u>positron emission</u>, and <u>gamma emission</u>.

18.5 A <u>radioactive decay series</u> is a sequence of decay steps that continues until a stable nucleus is reached.

18.6 The three naturally occurring radioactive series begin with uranium-238, uranium-235, and thorium-232.

18.7 The last member of the uranium-238 radioactive decay series is lead-206. It is formed in two different processes. First, thallium-206 decays by emitting a beta particle.

$$^{206}_{81}Tl \longrightarrow {}^{206}_{82}Pb + {}^{0}_{-1}e$$

The second process involves polonium-210. The nuclide decays by emitting an alpha particle

$$^{210}_{84}Po \longrightarrow {}^{206}_{82}Pb + {}^{4}_{2}He$$

Both processes produce lead-206 as a final, stable product.

18.8 Two naturally occurring nuclides that are not in any of the naturally occurring radioactive decay series are carbon-14 and potassium-40.

18.9 A <u>particle accelerator</u> is a device used to accelerate electrons, protons, and alpha particles to very high speeds. They are used in certain nuclear reactions to impart enough energy to the particle so that it can penetrate the nucleus of an element with a large atomic number.

18.10 The discovery of the transuranium elements has allowed scientists to extend the periodic chart up to element 109. All of the new elements are radioactive. Perhaps the most important of these elements is plutonium, atomic number 94, which is at the heart of the atomic weapons program. Also, as the seventh row (period) of the table is completed, it is hoped that new, stable elements might be discovered. These will be called the superheavy elements. Element with atomic number 118 will fall into the Noble gas group (VIII).

18.11 A <u>half-life</u> is the time required for half of the original sample of nuclei to decay. Thus, for cesium-137, with half-life of 30.2 years, after this length of time, one half of the original sample will have decayed away.

18.12 Technetium must be prepared in the laboratory because no isotopes of this element are members of a naturally occurring radioactive decay series. Uranium-238 is found on earth because its half-life, 4.51×10^9 years, is approximately the same as the age of the earth, so only one half-life for this isotope has elapsed. This means that approximately ½ of the uranium-238 that was present when the earth was formed is still on the earth. The longest lived isotope of technetium is much smaller (around 1 million years) so that any of this isotope that was present when the earth formed has long since decayed away and there is none left.

18.13 If an equal number of atoms of each nuclide in Table 18.2 are initially present, then after one billion years had elapsed, the nuclide that would be present in the largest amount would be the one with the longest half-life, or uranium-238. The nuclide that would be present in the smallest amount would be the one with the shortest half-life, or polonium-214. (As a side note, if there was only one sample with all of the nuclides present at the same time, then all of the nuclides from thorium-234 to thallium-206 have half-lives that are short enough compared to one billion years so that they would have all decayed to lead-206. Thus lead-206 would be the most prevalent nuclide in this case.)

18.14 Dating is a technique for determining the age of certain old objects that relies on the radioactive nuclides in the object. The number of half-lives that have elapsed since the object was formed multiplied by the value of one half-life is the age of the object.

18.15 In the absence of lead-208, we can assume that all of the lead-206 in the rock was once uranium-238. Since the ratio is 1:1, one half-life must have elapsed since the rock was formed. Thus, the age of the rock is about 4.5 billion years old.

18.16 Lead-208 is the most abundant isotope of lead and is in the rock as a result of ordinary processes rather than by radioactive decay. Since there is no uranium-238 nor lead-206 in the sample, this method cannot be used to date the rock in a reliable fashion. The age of the rock is unknown.

18.17 A Geiger counter is a device used to detect radioactivity, particularly alpha and beta particles and, in special circumstances, neutrons. A scintillation counter is an instrument that contains a phosphor, like a sodium iodide crystal, that emits flashes of light when it is struck with radiation. It can be used to detect gamma rays.

18.18 If radioactive phosphorus were administered to a living organism, it could be used as a radioactive tracer. Any radioactive nuclide can function as a tracer but it is desirable to have a nuclide with a relatively short half-life of a few days so that the organism is not subjected to radiation for a long period of time. Under these conditions, the phosphorus can be used to trace the pathway of the substance in the body. You could obtain information on where in the body the phosphorus is absorbed, how long it stays in the body, and the pathology of those parts of the body that absorb the phosphorus and use it as a nutrient.

18.19 The principal sources of radiation that the average person is exposed to include radon in the air, uranium and other radioactive elements in soil and rocks, and cosmic rays. Also, there is potassium-40 and carbon-14 in the body. About 55% of the average annual radiation comes from radon in the air.

18.20 A rem (from roentgen equivalent for man) is a unit for a radiation dose. Doses of 25 to 200 rems can cause a decrease in the number of white blood cells and doses up to 500 rems or more can cause death. Since the average annual exposure of a person in the United States to radiation from natural and human sources is about 0.360 rems, exposure to 0.10 rems of radioactivity would not have serious implications to you. According to the Nuclear Regulatory Commission, exposure to 0.360 rems corresponds to an annual risk of death from radiation-induced cancer of about 1800 in 10 million. Thus, exposure to 0.10 rems of radiation would have much less risk. However, I would still ask questions like "What type of radiation was I exposed to?", "Are there any short term effects to exposure to low doses of radiation?", "Are there any cumulative effects to exposure to low doses of radiation?", " Can the body repair the damage done by exposure to low doses of radiation?", "How confident is the Nuclear Regulatory Commission about its report of risk?", etc.

18.21 Radon itself is unreactive and can be inhaled without any chemical effects. However, radon-222 decays by emission of an alpha particle, which can be very dangerous when they are inside the body. Also, the product of the decay is polonium-218, which is a solid that will then be deposited in the lungs. It also decays by emitting an alpha particle. Many scientists believe that inhaled radon can cause lung cancer as a result.

18.22 Nuclear fission is a process in which nuclides with mass numbers greater than 50 split to form two new lighter and more stable nuclei. Nuclear fusion is a process in which nuclides with mass numbers less than 50 combine to give heavier, more stable nuclei.

18.23 Since carbon-12 is a nuclei with mass number less than 50, we would expect it to undergo fusion with other carbon-12 nuclei, rather than fission into smaller nuclei.

18.24 Fission of uranium-235 is induced by the absorption of neutrons, often from an outside source. As the nuclide splits by its many pathways, more neutrons are produced, which then are absorbed by other uranium-235 nuclides causing a chain reaction.

18.25 Critical mass is the minimum mass of a nuclide that is required to allow fission to become a self sustaining reaction. If the mass is larger than the critical mass, the chain reaction will continue rapidly and become supercritical and explode.

18.26 A nuclear power plant is not a potential atomic bomb waiting to explode. The fuel used is only enriched to about 3% uranium-235. This is not enough nuclides to reach critical mass and an uncontrolled chain reaction is not possible.

■ Solutions to Practice Problems

18.27 $^{87}_{37}Rb \longrightarrow ^{87}_{38}Sr + ^{0}_{-1}e$

18.28 $^{32}_{15}P \longrightarrow ^{32}_{16}S + ^{0}_{-1}e$

18.29 $^{232}_{90}Th \longrightarrow ^{228}_{88}Ra + ^{4}_{2}He$

18.30 $^{11}_{6}C \longrightarrow ^{11}_{5}B + ^{0}_{1}e$

18.31 If we let $^{A}_{Z}X$ represent the unknown product nuclide, we can write the decay process as

$$^{210}_{84}Po \longrightarrow ^{A}_{Z}X + ^{4}_{2}He$$

Since A and Z must be conserved we can write

A: 210 = A + 4,   or A = 206
Z: 84 = Z + 2,   or Z = 82

Therefore, the product nuclide must be lead-206. The nuclear equation can now be written as

$$^{210}_{84}Po \longrightarrow ^{206}_{82}Pb + ^{4}_{2}He$$

18.32 If we let $^{A}_{Z}X$ represent the unknown product nuclide, we can write the decay process as

$$^{227}_{89}Ac \longrightarrow ^{A}_{Z}X + ^{4}_{2}He$$

Since A and Z must be conserved we can write

A: 227 = A + 4,   or A = 223
Z: 89 = Z + 2,   or Z = 87

Therefore, the product nuclide must be francium-223. The nuclear equation can now be written as

$$^{227}_{89}Ac \longrightarrow ^{223}_{87}Fr + ^{4}_{2}He$$

18.33 If we let ^A_ZX represent the unknown product radiation, we can write the decay process as

$$^{18}_{9}F \longrightarrow ^{18}_{8}O + ^A_ZX$$

Since A and Z must be conserved, we can write

A: $18 = 18 + A$,   or $A = 0$
Z: $9 = 8 + Z$,   or $Z = 1$

The product radiation must have a mass number of zero and a charge of +1. This is a positron. Thus we can write the nuclear equation as

$$^{18}_{9}F \longrightarrow ^{18}_{8}O + ^{0}_{1}e$$

18.34 If we let ^A_ZX represent the unknown product radiation, we can write the decay process as

$$^{3}_{1}H \longrightarrow ^{3}_{2}He + ^A_ZX$$

Since A and Z must be conserved, we can write

A: $3 = 3 + A$,   or $A = 0$
Z: $1 = 2 + Z$,   or $Z = -1$

The product radiation must have a mass number of zero and charge of -1. This is a beta particle. Thus we can write the nuclear equation as

$$^{3}_{1}H \longrightarrow ^{3}_{2}He + ^{0}_{-1}e$$

18.35 If we let ^A_ZX represent the unknown product nuclide, we can write the nuclear equation as

$$^{6}_{3}Li + ^{1}_{0}n \longrightarrow ^{3}_{1}H + ^A_ZX$$

Since A and Z must be conserved, we can write

A: $6 + 1 = 3 + A$,   or $A = 4$
Z: $3 + 0 = 1 + Z$,   or $Z = 2$

Thus, the unknown nuclide must be $^{4}_{2}He$, an alpha particle.

18.36 If we let $^A_Z X$ represent the unknown product nuclide, we can write the nuclear equation as

$$^{27}_{13}Al + {}^3_1H \longrightarrow {}^{27}_{12}Mg + {}^A_Z X$$

Since A and Z must be conserved, we can write

A: 27 + 3 = 27 + A,   or A = 3
Z: 13 + 1 = 12 + Z,   or Z = 2

Thus, the unknown nuclide must be ^3_2He.

18.37 If we let $^A_Z X$ represent the unknown target nuclide, we can write the nuclear equation as

$$^A_Z X + {}^4_2He \longrightarrow {}^{242}_{96}Cm + {}^1_0n$$

Since A and Z must be conserved, we can write

A: A + 4 = 242 + 1,   or A = 239
Z: Z + 2 = 96 + 0,   or Z = 94

Thus, the unknown nuclide must be $^{239}_{94}Pu$.

18.38 If we let $^A_Z X$ represent the unknown target nuclide, we can write the nuclear equation as

$$^A_Z X + {}^4_2He \longrightarrow {}^{245}_{98}Cf + {}^1_0n$$

Since A and Z must be conserved, we can write

A: A + 4 = 245 + 1,   or A = 242
Z: Z + 2 = 98 + 0,   or Z = 96

Thus, the unknown nuclide must be $^{242}_{96}Cm$.

18.39 After three half-lives for <u>any</u> nuclide, there will be ½ x ½ x ½ = ⅛ of the original sample remaining.

18.40 After six half-lives for <u>any</u> nuclide, there will be ½ x ½ x ½ x ½ x ½ x ½ = 1/64 of the original sample remaining.

282 ■ CHAPTER 18

18.41 First, we must determine the approximate number of half-lives that have elapsed. Two weeks is 8 days, so we can write

$$8 \text{ days} \times \frac{1 \text{ half-life}}{2.69 \text{ days}} = 2.97 \text{ half-lives}$$

If we round this off to approximately 3 half-lives, then ½ x ½ x ½ = ⅛ of the original sample will remain, or

$$\frac{1}{8} \times 0.00100 \text{ g} = 1.25 \times 10^{-4} \text{ g} = 1 \times 10^{-4} \text{ g}$$

18.42 First, we must determine the approximate number of half-lives that have elapsed. For 200 years, we can write

$$200 \text{ years} \times \frac{1 \text{ half-life}}{2.05 \text{ years}} = 97.6 \text{ half-lives}$$

If we round this off to 98 half-lives, the amount of material that is left is

$$(\tfrac{1}{2})^{98} \times 1{,}000{,}000{,}000 \text{ atoms} = 3.2 \times 10^{-21} \text{ atoms} = 3 \times 10^{-21} \text{ atoms}$$

Since this is less than one atom, all of the sample is all gone.

18.43 First, we must determine the number of half-lives that have elapsed. The disintegrations per minute will drop from 16 to 8 in one half-life. Thus, the cyprus beam must be approximately one half-life, or 5700 years, old. The tree was felled approximately

$$2000 - 5700 = -3700 \text{ AD} \quad \text{or} \quad 3700 \text{ BC}$$

18.44 First, we must determine the approximate number of half-lives that have elapsed. The disintegrations per minute will drop from 16 to 2 in three half-lives. Thus, the bones must be approximately three half-lives old, or 3 x 5700 years = 17,100 years old. The animal died approximately

$$2000 - 17{,}100 = -15{,}100 \text{ AD} \quad \text{or} \quad 15{,}100 \text{ BC}$$

18.45 First, we must determine the approximate number of half-lives that have elapsed. The disintegrations per minute will drop from 16 to 4 in two half-lives. Thus, the artifact is approximately two half-lives, or 2 x 5700 years = 11,400 years old.

18.46 First, we must determine the approximate number of half-lives that have elapsed. The disintegrations per minute will drop from 16 to 1 in four half-lives. We can use four half-lives as a rouge estimate for the calculation. Thus, the artifact is approximately 4 x 5700 years, or 22,800 years old.

■ Solutions to Additional Problems

18.47 If we let A be the unknown mass number of the astatine nucleus, we can write

$$^{209}_{83}Bi + ^{4}_{2}He \longrightarrow ^{A}_{85}At + 2\,^{1}_{0}n$$

Since A must be conserved, we can write

A: $209 + 4 = A + 2 \times 1$,   or A = 211

Thus, the product nuclide must be astatine-211. The nuclear equation for this reaction is

$$^{209}_{83}Bi + ^{4}_{2}He \longrightarrow ^{211}_{85}At + 2\,^{1}_{0}n$$

18.48 If we let A be the unknown mass number of the polonium nucleus, we can write

$$^{209}_{83}Bi + ^{2}_{1}H \longrightarrow ^{A}_{84}Po + ^{1}_{0}n$$

Since A must be conserved, we can write

A: $209 + 2 = A + 1$,   or A = 210

Thus, the product nuclide must be polonium-210. The nuclear equation for this reaction is

$$^{209}_{83}Bi + ^{2}_{1}H \longrightarrow ^{210}_{84}Po + ^{1}_{0}n$$

18.49 If we let $^{A}_{Z}X$ represent the unknown nuclide, then to conserve A and Z, we can write

A: $238 + 12 = A + 4 \times 1$,   or A = 246
Z: $92 + 6 = Z + 4 \times 0$,   or Z = 98

The unknown nuclide must be californium-246. The nuclear equation for this reaction is

$$^{238}_{92}U + ^{12}_{6}C \longrightarrow ^{246}_{98}Cf + 4\,^{1}_{0}n$$

18.50 If we let ^A_ZX represent the unknown nuclide, then to conserve A and Z, we can write

A: $246 + 12 = A + 4 \times 1$,   or A = 254
Z: $96 + 6 = Z + 4 \times 0$,   or Z = 102

The unknown nuclide must be nobelium-254. The nuclear equation for this reaction is

$$^{246}_{96}Cm + ^{12}_{6}C \longrightarrow ^{254}_{102}No + 4\,^{1}_{0}n$$

19. ORGANIC CHEMISTRY

■ Solutions to Exercises

19.1 a. Structural formulas of butane (n-butane) and isobutane (2-methylpropane).

b. Structural formulas of 1-butanol, 2-butanol, 2-methyl-1-propanol, and 2-methyl-2-propanol.

286 ■ CHAPTER 19

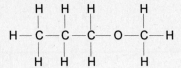

19.2 a. The longest carbon chain is five atoms, so the parent name is pentane. There are two methyl groups attached to the parent chain.

$$CH_3\,^3CH\,^2CH\,^1CH_3$$
with CH_3 above ^2CH and $^4CH_2\,^5CH_3$ below ^3CH

The name is 2,3-dimethylpentane.

b. The longest carbon chain is four atoms, so the parent name is butane. There are two methyl groups attached.

$$^4CH_3\,^3CH\,^2CH\,^1CH_3$$
with CH_3 above ^2CH and CH_3 below ^3CH

The name is 2,3-dimethylbutane.

19.3 3-methylpentane has a parent chain of five carbons and one methyl branch on carbon 3. Fill in hydrogens to satisfy four bonds for each carbon. The final structure is

```
              CH₃
    H   H     |    H   H
    |   |     |    |   |
H—C—C—C—C—C—H
    |   |     |    |   |
    H   H     H    H   H
```

19.4 a. Propane is $CH_3CH_2CH_3$; The molecular formula is C_3H_8. The balanced equation is

$$C_3H_8(g) + 5\,O_2(g) \longrightarrow 3\,CO_2(g) + 4\,H_2O(g)$$

b. There are many possibilities for the reaction of chlorine with butane, C_4H_{10}. Some are

$$CH_3CH_2CH_2CH_3 + Cl_2 \longrightarrow CH_3CH_2CH_2CH_2Cl + HCl$$

$$CH_3CH_2CH_2CH_3 + Cl_2 \longrightarrow CH_3CH_2CHClCH_3 + HCl$$

19.5 a. The longest chain is six carbons, and it has a triple bond.

$$^6CH_3\,^5CH_2\,^4CH_2\,^3C\equiv{}^2C\,^1CH_3$$

The name of the compound is 2-hexyne.

b. The longest chain is seven carbons. Number from the end nearest the double bond.

$$\begin{array}{c} CH_3 \\ | \\ ^1CH_3\,^2CH\,^3CH={}^4CH\,^5CH_2\,^6CH_2\,^7CH_3 \end{array}$$

The name of the compound is 2-methyl-3-heptene.

288 ■ CHAPTER 19

19.6   a. m-chloronitrobenzene   b. 1,2,4-tribromobenzene

   c. p-bromotoluene   d. nitrobenzene

19.7 The longest chain containing the OH group is six so the parent name is hexanol. There are two methyl groups. Begin numbering from the end closest to the OH group.

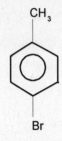

The name of the compound is 3,5-dimethyl-3-hexanol.

19.8 a. The compound is an aldehyde; the numbering of the parent chain is

The name of the compound is 4-methylhexanal.

b. The compound is a ketone; The numbering of the parent chain is

$^5CH_3\,^4CH_2\,^3CH_2\,^2CO\,^1CH_3$

The name of the compound is 2-pentanone.

■ Answers to Questions to Test Your Reading

19.1 Organic chemicals around the house may include aspirin, rubbing alcohol, nail polish remover, acetone, clothing made of polyester or nylon, etc.

19.2 The present definition of <u>organic chemistry</u> is the chemistry of carbon compounds. The original definition of organic compounds were substances obtained from plant and animal materials.

19.3 Carbon can bond in the following four ways

methane, CH_4 formaldehyde, H_2CO carbon dioxide, CO_2 acetylene, C_2H_2

19.4 <u>Isomers</u> are compounds that have the same molecular formula but different structural formulas. An example would be C_2H_6O, where the two structures for ethanol and diethyl ether are isomers.

19.5 The condensed structural formula of butanol would be $CH_3CH_2CH_2CH_2OH$.

19.6 The first four members of the alkane series are methane, CH_4, ethane, C_2H_6, propane, C_3H_8, and butane, C_4H_{10}.

19.7 A straight-chain alkane of seven carbons would be heptane, C_7H_{16}.

19.8 The alkyl group obtained from propane is propyl. (see Table 19.2)

290 ■ CHAPTER 19

19.9 The major chemical sources of alkanes are natural gas and petroleum. These materials occur naturally as the end products of the decay of organic material in the earth.

The major uses of natural gas and petroleum alkanes are as starting materials for the production of organic chemicals. Natural gas is used for home and industrial heating. Gasoline is the principal product of petroleum. Other uses of petroleum include various fuels, solvents, oils and lubricants (see Table 19.3).

19.10 Petroleum refining is a chemical process for petroleum to produce a fuel suitable for modern automobiles.

19.11 Straight-chain alkanes are unsuitable as gasoline because they tend to preignite causing the engine to "knock", which causes undue wear on an engine. They have low octane ratings.

19.12 The combustion of methane is given by the reaction

$$CH_4(g) + 2 O_2(g) \longrightarrow CO_2(g) + 2 H_2O(g)$$

The reaction of methane with chlorine is given by

$$CH_4 + Cl_2 \longrightarrow CH_3Cl + HCl$$

19.13 Two alkenes are ethylene, C_2H_4, and propylene, C_3H_6. Their structures are

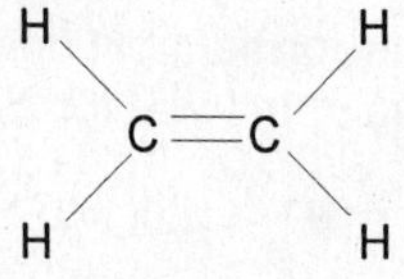

[structures of ethylene and propylene shown] and [structure shown]

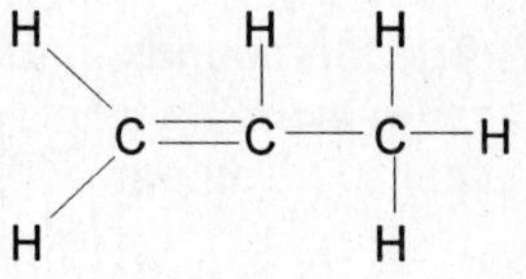

An example of an alkyne is acetylene, C_2H_2, whose structure is

$$H-C\equiv C-H$$

19.14 Ethylene (and propylene) are present in small quantities in natural gas and petroleum. In the chemical industry it is produced from ethane by heating. It is used to prepare plastics like polyethylene.

19.15 The reaction for the addition of bromine to ethylene is

$$CH_2=CH_2 + Br_2 \longrightarrow \underset{\underset{Br}{|}}{CH_2}\underset{\underset{Br}{|}}{CH_2}$$

The name of the product is 1,2-dibromoethane. This reaction can be used as a test for multiple bonds. The red-brown color of bromine disappears as it reacts with the multiple bond.

19.16 A <u>polymer</u> is a very large molecule consisting of many repeating units of low molecular weight. Polyethylene is an example. It is a chain of repeating C_2H_4 units. The <u>monomer</u> is a compound used to prepare a polymer and gives rise to the repeating unit. Polyethylene means "many ethylene monomer units".

19.17 Polypropylene looks like

$$-\underset{\underset{H}{|}}{\overset{\overset{CH_3}{|}}{C}}-\underset{\underset{H}{|}}{\overset{\overset{H}{|}}{C}}-\underset{\underset{H}{|}}{\overset{\overset{CH_3}{|}}{C}}-\underset{\underset{H}{|}}{\overset{\overset{H}{|}}{C}}-\underset{\underset{H}{|}}{\overset{\overset{CH_3}{|}}{C}}-\underset{\underset{H}{|}}{\overset{\overset{H}{|}}{C}}-$$

19.18 The resonance description of benzene is

[benzene resonance structures with double-headed arrow between two Kekulé structures]

The meaning of the double arrow is that the extra pairs of electrons represented by double bonds are actually delocalized over the benzene ring, rather than localized between two carbon atoms as shown by any single formula. One pair of electrons in each double bond occupies a molecular orbital encompassing all six carbon atoms.

19.19 <u>Para</u> is a prefix used in a disubstituted benzene ring where the substituents are on the carbon atoms on opposite sides of the ring.

292 ■ CHAPTER 19

19.20 A <u>functional group</u> is a reactive position of a molecule that undergoes predictable reactions. Some examples are alkenes (double bonds), alcohols (ROH), ethers (ROR'), etc. (see Table 19.4)

19.21 An <u>alcohol</u> is a compound ROH that contains an OH group bonded to a tetrahedral carbon atom (an alkyl group R). Some examples are methanol, CH_3OH, and ethanol CH_3CH_2OH. An <u>ether</u> is a compound ROR' that contains an O atom bonded to two hydrocarbon groups, R and R'. An example is diethyl ether, $CH_3CH_2OCH_2CH_3$, used as an anesthetic.

19.22 Methanol is produced from the reaction of carbon monoxide and hydrogen, using a metal oxide catalyst

$$CO\ (g)\ +\ 2\ H_2\ (g)\ \xrightarrow[400\,°C,\ 150\ atm]{ZnO\ \cdot\ Cr_2O_3}\ CH_3OH\ (g)$$

Methanol is a common solvent for organic materials, such as shellac and varnish. It is the main ingredient in automobile windshield washer compounds, and as a fuel.

19.23 Ethanol is prepared from ethylene by reacting the ethylene with steam in the presence of an acid catalyst (such as sulfuric acid).

$$CH_2{=}CH_2\ +\ HOH\ \xrightarrow[heat]{H^+}\ CH_3CH_2OH$$

19.24 1-propanol will be oxidized by an acidic dichromate ion solution into propanoic acid as a final product.

```
    H   H   O
    |   |   ||
H—C—C—C—O—H
    |   |
    H   H
```

19.25 a. 3-methyl-3-heptanol is a tertiary alcohol and is unreactive towards acidic dichromate ion solutions.

b. 2-methyl-3-heptanol is a secondary alcohol and will react to form a ketone, 2-methyl-3-heptanone.

19.26 a. aldehyde   b. alcohol   c. ketone

19.27 Formaldehyde, HCHO, is a gas with a characteristic irritating odor. An aqueous solution containing methanol is called formalin and is used as a preservative and disinfectant. A commercial use is in the manufacture of polymers.

Acetone, CH_3COCH_3, is a flammable, volatile liquid with a slight aromatic odor. It is a solvent for certain resins and acetate plastics. In industry it is used in the manufacture of methyl methacrylate, which is made into polymers like Lucite and Plexiglas.

19.28 You do not need a number for propanone, because the carbonyl group of the ketone can only be on the number two carbon atom; otherwise, it would be an aldehyde.

19.29 Acetic acid and its corresponding ionization reaction is given by

$$CH_3COOH + H_2O \longrightarrow CH_3COO^- + H_3O^+$$

← carboxylic acid group

19.30 Acetic acid + methanol ⟶ methyl acetate + water

$$CH_3C(=O)OH + HOCH_3 \xrightarrow{H^+} CH_3C(=O)-O-CH_3 + H_2O$$

19.31 Aniline + hydronium ion ⟶ phenylammonium ion + water

$$C_6H_5NH_2 + H_3O^+ \longrightarrow [C_6H_5NH_3]^+ + H_2O$$

19.32 Acetic acid + methylamine —> methylacetamide + water

$$CH_3-\overset{\overset{O}{\|}}{C}-OH \ + \ H-\underset{\underset{H}{|}}{\overset{\overset{CH_3}{|}}{N}} \ \xrightarrow{\Delta} \ CH_3-\overset{\overset{O}{\|}}{C}-\underset{\underset{H}{|}}{\overset{\overset{CH_3}{|}}{N}} \ + \ H_2O$$

■ Solutions to Practice Problems

19.33a. Chloroform has four bonding groups around the central atom and is tetrahedral.

   b. Formaldehyde has three bonding groups around the central atom and is trigonal planar.

   c. Hydrogen cyanide has two bonding groups around the central atom and is linear.

19.34a. Phosgene has three bonding groups around the central atom and is trigonal planar.

   b. Methyl bromide has four bonding groups around the central atom and is tetrahedral.

   c. Cyanogen chloride has two bonding groups around the central atom and is linear.

19.35a.

b.

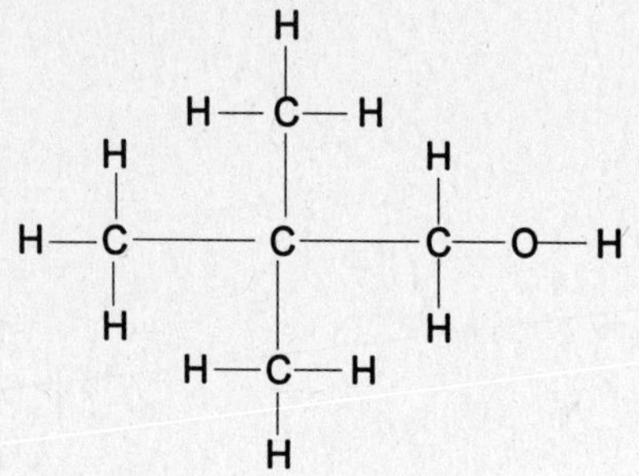

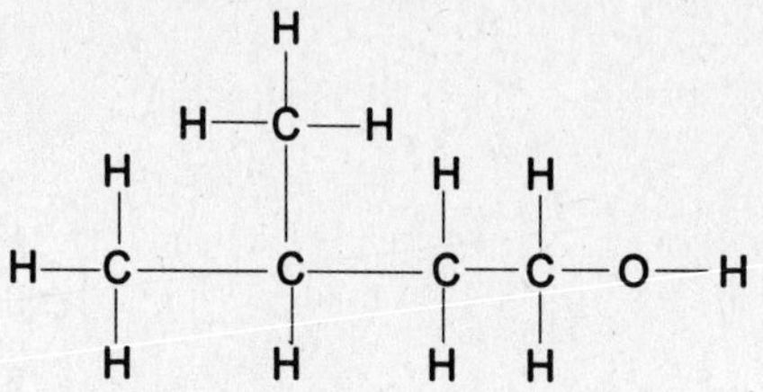

296 ■ CHAPTER 19

19.36a.

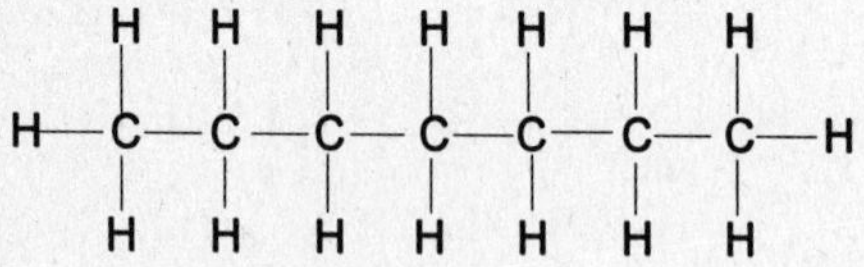

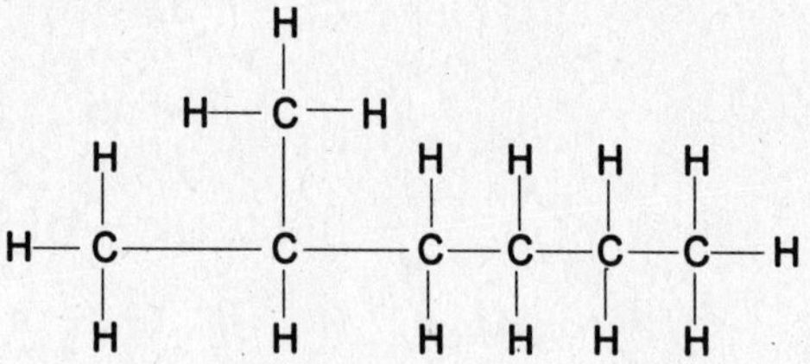

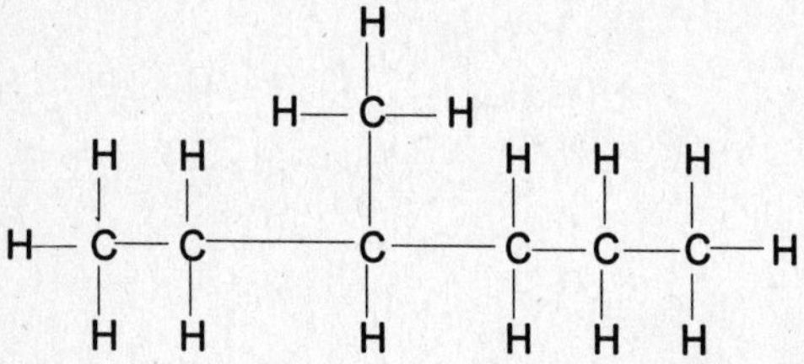

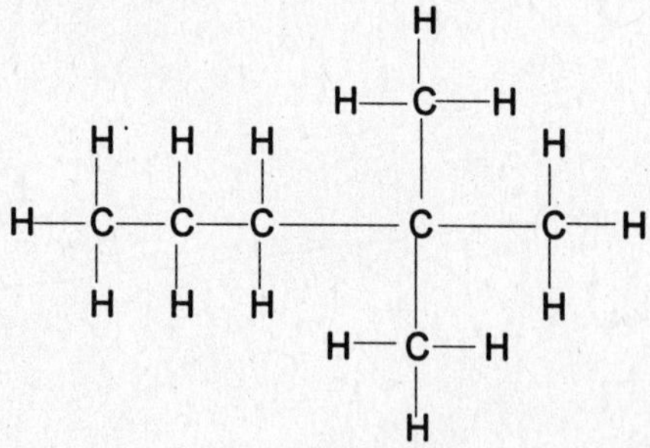

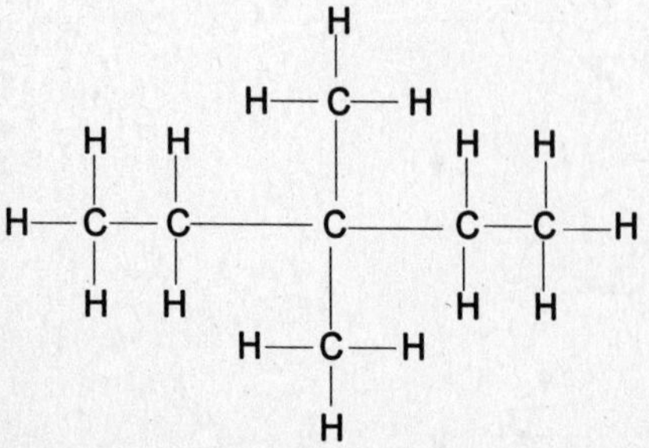

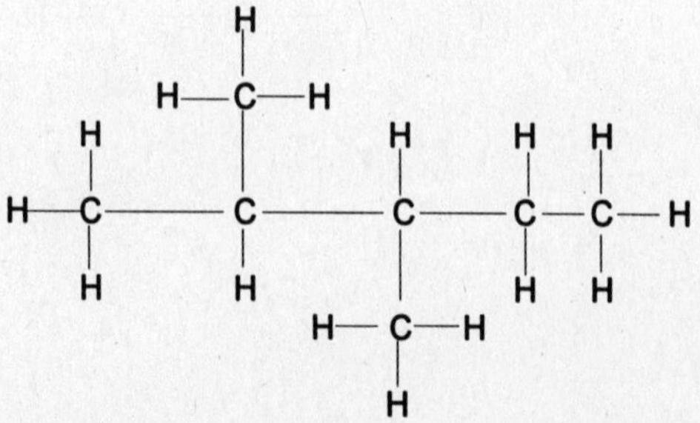

There are three other structures that are not given here.

b.

19.37a.

b.

19.38a.

b.

$$\text{H}-\underset{\underset{\text{H}}{|}}{\overset{\overset{\text{H}}{|}}{\text{C}}}-\underset{\underset{\text{H}}{|}}{\overset{\overset{\text{H}}{|}}{\text{C}}}-\underset{\underset{\text{H}}{|}}{\overset{\overset{\text{Cl}}{|}}{\text{C}}}-\text{Cl}$$

$$\text{H}-\underset{\underset{\text{H}}{|}}{\overset{\overset{\text{H}}{|}}{\text{C}}}-\underset{\underset{\text{H}}{|}}{\overset{\overset{\text{Cl}}{|}}{\text{C}}}-\underset{\underset{\text{H}}{|}}{\overset{\overset{\text{H}}{|}}{\text{C}}}-\text{Cl}$$

$$\text{H}-\underset{\underset{\text{H}}{|}}{\overset{\overset{\text{Cl}}{|}}{\text{C}}}-\underset{\underset{\text{H}}{|}}{\overset{\overset{\text{H}}{|}}{\text{C}}}-\underset{\underset{\text{H}}{|}}{\overset{\overset{\text{Cl}}{|}}{\text{C}}}-\text{H}$$

$$\text{H}-\underset{\underset{\text{H}}{|}}{\overset{\overset{\text{H}}{|}}{\text{C}}}-\underset{\underset{\text{Cl}}{|}}{\overset{\overset{\text{Cl}}{|}}{\text{C}}}-\underset{\underset{\text{H}}{|}}{\overset{\overset{\text{H}}{|}}{\text{C}}}-\text{H}$$

19.39 a. With a CH₃ branch on the middle carbon of a three-carbon chain ending in C—O—H.

b. H—CH₂—O—C(CH₃)H—CH₂—CH₃ type structure with CH₃ branch.

19.40 a. H—O—CH₂—CH₂—O—H

b. H—CH₂—O—C(CH₃)H—CH₂—CH₂—O—H type structure with CH₃ branch.

19.41 The molecular formula is C_8H_{18}. The condensed structural formula is

$CH_3CH_2CH_2CH_2CH_2CH_2CH_2CH_3$

The full structural formula is

$$\text{H}-\overset{\overset{\text{H}}{|}}{\underset{\underset{\text{H}}{|}}{\text{C}}}-\overset{\overset{\text{H}}{|}}{\underset{\underset{\text{H}}{|}}{\text{C}}}-\overset{\overset{\text{H}}{|}}{\underset{\underset{\text{H}}{|}}{\text{C}}}-\overset{\overset{\text{H}}{|}}{\underset{\underset{\text{H}}{|}}{\text{C}}}-\overset{\overset{\text{H}}{|}}{\underset{\underset{\text{H}}{|}}{\text{C}}}-\overset{\overset{\text{H}}{|}}{\underset{\underset{\text{H}}{|}}{\text{C}}}-\overset{\overset{\text{H}}{|}}{\underset{\underset{\text{H}}{|}}{\text{C}}}-\overset{\overset{\text{H}}{|}}{\underset{\underset{\text{H}}{|}}{\text{C}}}-\text{H}$$

ORGANIC CHEMISTRY ■ 299

19.42 The molecular formula is C_9H_{20}. The condensed structural formula is

$CH_3CH_2CH_2CH_2CH_2CH_2CH_2CH_2CH_3$

The full structural formula is

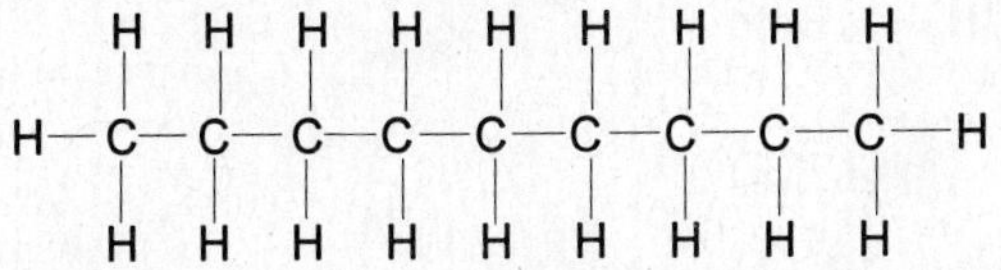

```
    H  H  H  H  H  H  H  H  H
    |  |  |  |  |  |  |  |  |
H — C— C— C— C— C— C— C— C— C —H
    |  |  |  |  |  |  |  |  |
    H  H  H  H  H  H  H  H  H
```

19.43   a. 3-methylhexane   b. 2,2-dimethylbutane   c. 3-ethylhexane

19.44   a. 2,3-dimethylhexane   b. 4-ethyl-4-methylheptane

       c. 3-ethyl-3-methylhexane

19.45

a. 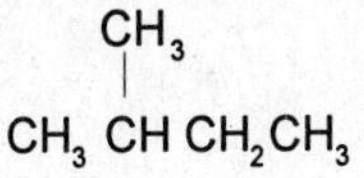
$$CH_3 \underset{\underset{CH_3}{|}}{CH} CH_2 CH_3$$

b. 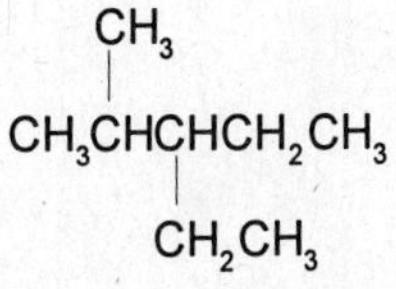
$$CH_3 \underset{\underset{CH_3}{|}}{CH} CHCH_2 CH_3$$
$$|$$
$$CH_2CH_3$$

19.46

a. 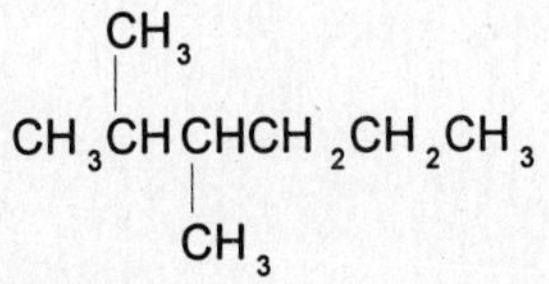
$$CH_3 CHCHCH_2 CH_3$$
with CH_3 above first CH and CH_3 below second CH

b. 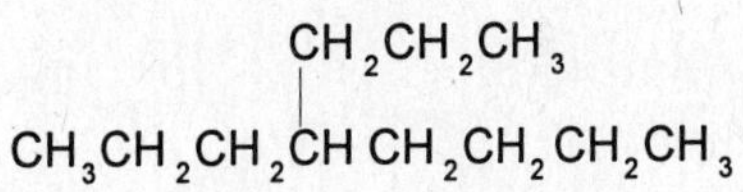
$$CH_3CH_2CH_2 \underset{\underset{}{|}}{CH} CH_2CH_2CH_2CH_3$$
with $CH_2CH_2CH_3$ above

19.47 C_7H_{16} + 11 O_2 ⟶ 7 CO_2 + 8 H_2O

19.48 2,2,4-trimethylpentane has the chemical formula C_8H_{18}, and is an isomer of octane. The reaction is

2 C_8H_{18} + 25 O_2 ⟶ 16 CO_2 + 18 H_2O

19.49 $CH_3CH_2CH_2CH_2CH_3$ + Cl_2 ⟶ $CH_3CH_2CH_2CH_2CH_2Cl$ + HCl

19.50 $CH_3 \underset{\underset{}{|}}{CH} CH_2CH_3$ with CH_3 above + Br_2 ⟶ $CH_3 \underset{\underset{}{|}}{CH} CH_2CH_2Br$ with CH_3 above + HBr

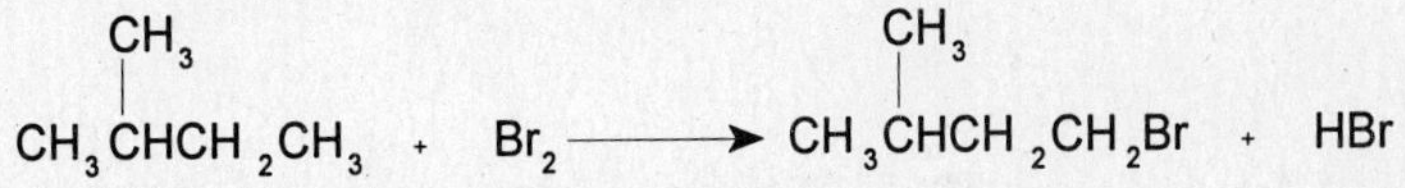

19.51 a. 4-methyl-1-pentene b. 3-methylbutyne

19.52 a. 3-ethyl-1-pentene b. 2,5-dimethyl-3-hexyne

19.53 $CH_3CH=CH_2 + Br_2 \longrightarrow CH_3\underset{\underset{Br}{|}}{C}HCH_2Br$

19.54 $CH_2=CHCH_2CH_3 + Cl_2 \longrightarrow CH_2\underset{\underset{Cl}{|}}{C}H\underset{\underset{}{}}{C}H_2CH_3$ (Cl on second carbon as well)

19.55

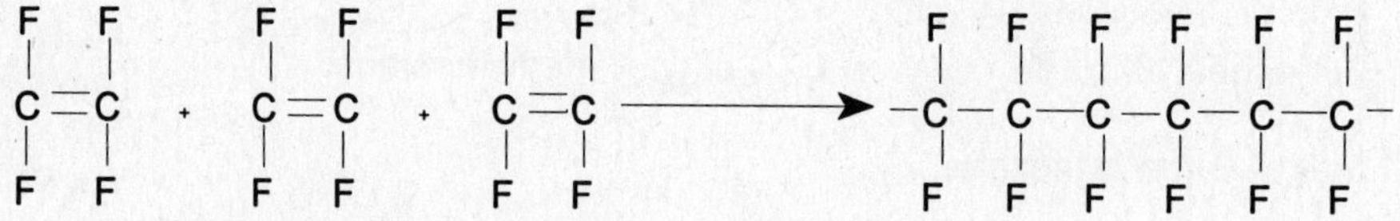

$$\underset{F}{\overset{F}{|}}C=\underset{F}{\overset{F}{|}}C \;+\; \underset{F}{\overset{F}{|}}C=\underset{F}{\overset{F}{|}}C \;+\; \underset{F}{\overset{F}{|}}C=\underset{F}{\overset{F}{|}}C \longrightarrow -\underset{F}{\overset{F}{|}}C-\underset{F}{\overset{F}{|}}C-\underset{F}{\overset{F}{|}}C-\underset{F}{\overset{F}{|}}C-\underset{F}{\overset{F}{|}}C-\underset{F}{\overset{F}{|}}C-$$

19.56

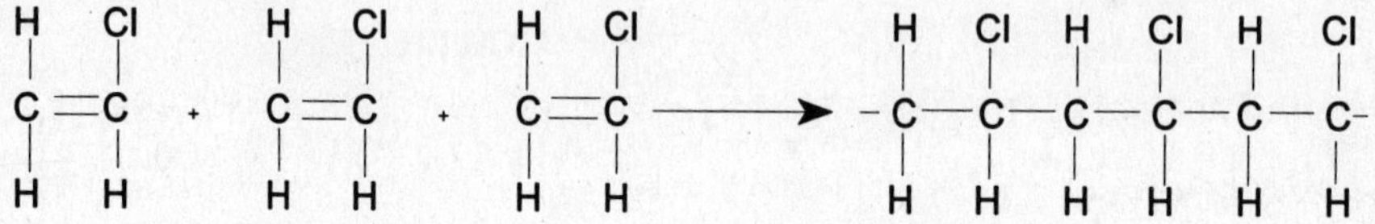

$$\underset{H}{\overset{H}{|}}C=\underset{H}{\overset{Cl}{|}}C \;+\; \underset{H}{\overset{H}{|}}C=\underset{H}{\overset{Cl}{|}}C \;+\; \underset{H}{\overset{H}{|}}C=\underset{H}{\overset{Cl}{|}}C \longrightarrow -\underset{H}{\overset{H}{|}}C-\underset{H}{\overset{Cl}{|}}C-\underset{H}{\overset{H}{|}}C-\underset{H}{\overset{Cl}{|}}C-\underset{H}{\overset{H}{|}}C-\underset{H}{\overset{Cl}{|}}C-$$

19.57 (toluene resonance structures ↔)

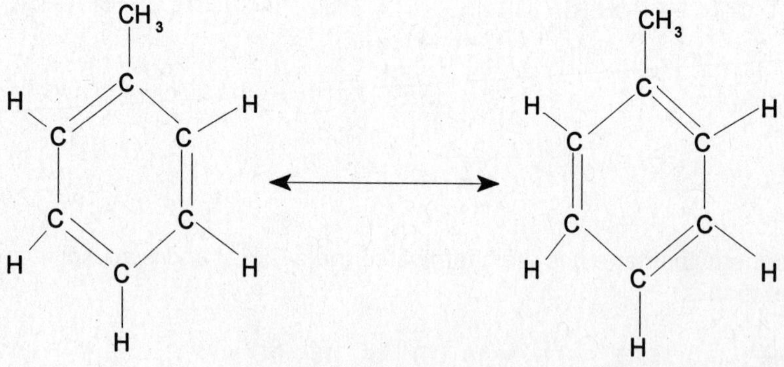

19.58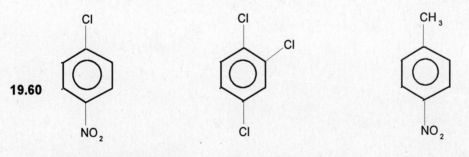

19.59 a. m-diethylbenzene b. o-bromotoluene c. 1,3,5-tribromobenzene

19.60 a. p-chloronitrobenzene b. 1,2,4-trichlorobenzene c. p-nitrotoluene

19.61 a. m-dibromobenzene b. 1,2,3-tribromobenzene

19.62 a. m-chlorotoluene b. p-diethylbenzene

19.63 a. tertiary b. primary c. secondary

19.64 a. secondary b. tertiary c. primary

19.65 3-methyl-2-pentanol

19.66 2,3-dimethyl-2-pentanol

19.67a. $CH_3CH_2CH_2CH_2OH + 6\ O_2 \longrightarrow 4\ CO_2 + 5\ H_2O$

b. $3\ CH_3CH_2CH_2CH_2OH + Cr_2O_7^{2-} + 8\ H^+ \longrightarrow$

$$3\ CH_3CH_2CH_2\overset{\overset{O}{\|}}{C}-H + 2\ Cr^{3+} + 7\ H_2O$$

c. $3\ CH_3\overset{\overset{OH}{|}}{CH}CH_2CH_2CH_3 + Cr_2O_7^{2-} + 8\ H^+ \longrightarrow$

$$3\ CH_3\overset{\overset{O}{\|}}{C}CH_2CH_2CH_3 + 2\ Cr^{3+} + 7\ H_2O$$

d. no reaction - tertiary alcohols do not react with acidic dichromate solutions.

19.68a. $2\ CH_3\underset{\underset{CH_3}{|}}{\overset{\overset{OH}{|}}{C}}CH_2CH_3 + 15\ O_2 \longrightarrow 10\ CO_2 + 12\ H_2O$

b. $3\ CH_3\underset{\underset{CH_3}{|}}{\overset{\overset{OH}{|}}{C}H}CHCH_3 + Cr_2O_7^{2-} + 8\ H^+ \longrightarrow 3\ CH_3\underset{\underset{CH_3}{|}}{\overset{\overset{O}{\|}}{C}}CHCH_3 + 2\ Cr^{3+} + 7\ H_2O$

c. no reaction - tertiary alcohols do not react with acidic dichromate solutions.

d. $3\ CH_3CH_2CH_2CH_2CH_2OH + Cr_2O_7^{2-} + 8\ H^+ \longrightarrow$

$$3\ CH_3CH_2CH_2CH_2\overset{\overset{O}{\|}}{C}-H + 2\ Cr^{3+} + 7\ H_2O$$

19.69 a. dimethyl ether, or methoxymethane b. ethylpropyl ether, or ethoxypropane

19.70 a. methyl propyl ether, or methoxypropane b. dipropyl ether, or propoxypropane

19.71 a. 2-methylbutanal b. 3-methyl-2-butanone

19.72 a. 3,3-dimethylbutanal b. 2-methyl-3-pentanone

19.73 a.
$$CH_3CH_2\underset{\underset{CH_3}{|}}{CH}\underset{\underset{CH_3}{|}}{CH}-\overset{\overset{O}{\|}}{C}-H$$

b.
$$CH_3\overset{\overset{O}{\|}}{C}-\underset{\underset{CH_3}{|}}{\overset{\overset{CH_3}{|}}{C}}-CH_2CH_3$$

19.74 a.
$$CH_3CH_2CH_2\underset{\underset{CH_2CH_3}{|}}{CH}CH_2-\overset{\overset{O}{\|}}{C}-H$$

b.
$$CH_3\overset{\overset{O}{\|}}{C}-\underset{\underset{CH_2CH_3}{|}}{CH}CH_2CH_2CH_3$$

19.75 2,2-dimethylpropanoic acid

19.76 3,4-dimethylpentanoic acid

19.77
$$CH_3\underset{\underset{CH_3}{|}}{CH}-\overset{\overset{O}{\|}}{C}-OH$$

19.78
$$CH_3CH_2\underset{\underset{CH_3}{|}}{\overset{\overset{CH_3}{|}}{C}}-\overset{\overset{O}{\|}}{C}-OH$$

19.79
$$CH_3CH_2CH_2\overset{\overset{O}{\|}}{C}-OH + CH_3\underset{\underset{}{}}{\overset{\overset{OH}{|}}{CH}}CH_3 \xrightarrow{H^+} CH_3CH_2CH_2\overset{\overset{O}{\|}}{C}-O-\underset{\underset{CH_3}{|}}{\overset{\overset{CH_3}{|}}{CH}} + H_2O$$

304 ■ CHAPTER 19

19.80 $CH_3CH_2\overset{\overset{O}{\|}}{C}-OH$ + $HO-\overset{\overset{CH_3}{|}}{C}HCH_2CH_3$ $\xrightarrow{H^+}$

$CH_3CH_2\overset{\overset{O}{\|}}{C}-O-\overset{\overset{CH_3}{|}}{C}HCH_2CH_3$ + H_2O

19.81 $CH_3CH_2CH_2CH_2\overset{\overset{H}{|}}{N}-H$ + H_3O^+ $\longrightarrow$ $\left[CH_3CH_2CH_2CH_2\overset{\overset{H}{|}}{\underset{H}{N}}-H\right]^+$ + H_2O

19.82 $H-\overset{\overset{CH_2CH_3}{|}}{\underset{CH_2CH_3}{N}}$ + H_3O^+ $\longrightarrow$ $\left[H-\overset{\overset{CH_2CH_3}{|}}{\underset{CH_2CH_3}{N}}-H\right]^+$ + H_2O

19.83 $H-\overset{\overset{H}{|}}{N}-H$ + $HO-\overset{\overset{O}{\|}}{C}CH_2CH_3$ $\xrightarrow{\Delta}$ $CH_3CH_2\overset{\overset{O}{\|}}{C}-NH_2$ + H_2O

19.84 $H-\overset{\overset{CH_3}{|}}{N}-H$ + $HO-\overset{\overset{O}{\|}}{C}CH_2CH_3$ $\xrightarrow{\Delta}$ $CH_3CH_2\overset{\overset{O}{\|}}{C}-\overset{\overset{}{\underset{CH_3}{N}}}{-}H$ + H_2O

■ Solutions to Additional Problems

19.85 a. $H-\overset{\overset{H}{|}}{\underset{H}{C}}-\overset{\overset{H}{|}}{\underset{H}{C}}-OH$ \qquad b. $H-\overset{\overset{H}{|}}{\underset{H}{C}}-\overset{\overset{O}{\|}}{C}-H$

c. 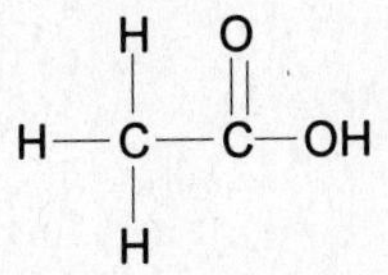 structure of acetic acid (H-CH2-C(=O)-OH)

d. 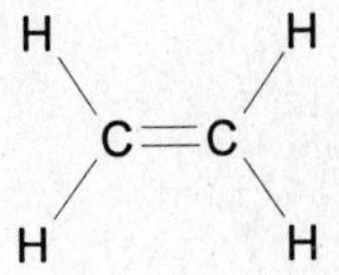 structure of ethene (H2C=CH2)

19.86 a. 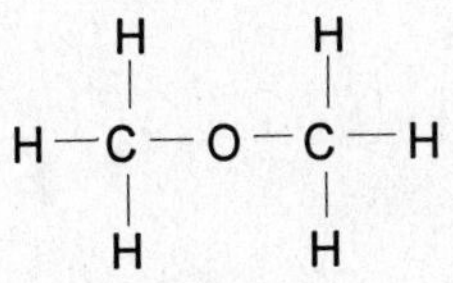 H-CH2-O-CH2-H (dimethyl ether)

b. 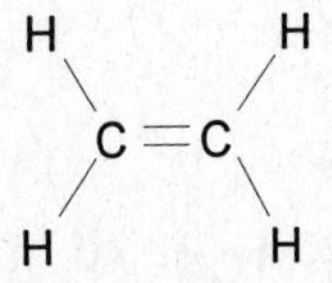 H2C=CH2

c. 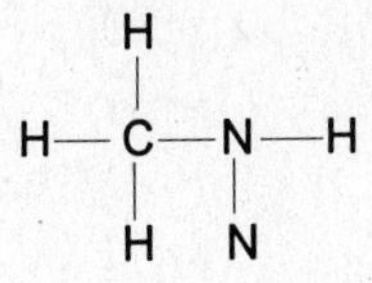 H-CH2-N(H)(H) with N shown

d. 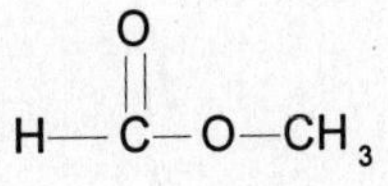 H-C(=O)-O-CH3

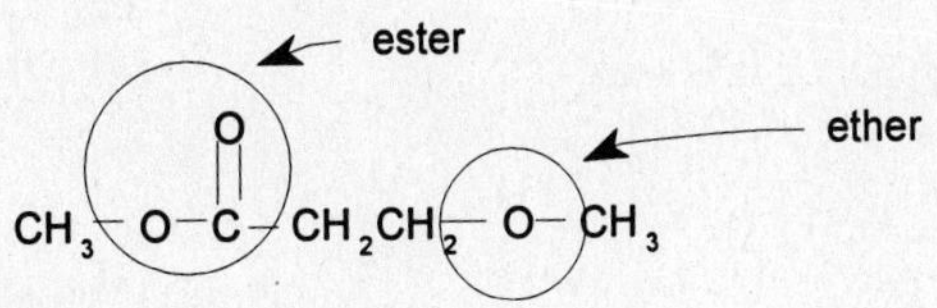

19.87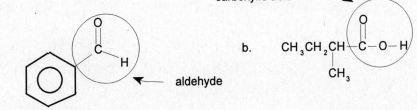

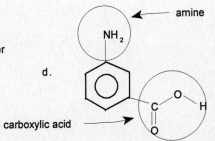

19.88 a. 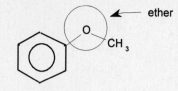 ether

b. 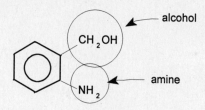 alcohol, amine

c. CH₃CH₂CH(NH₂)—C(=O)—OH — carboxylic acid, amine

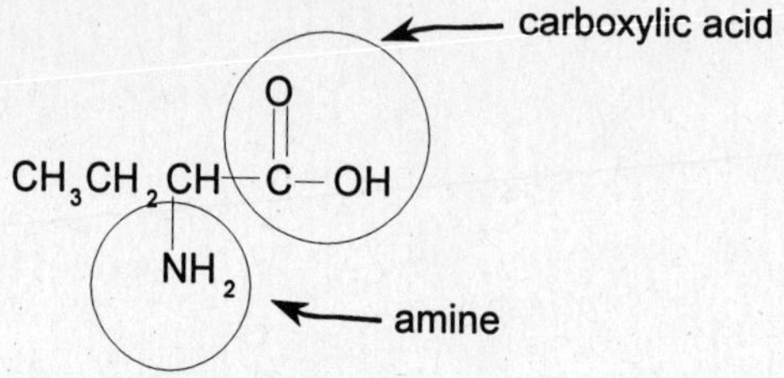

d. 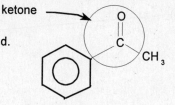 ketone

19.89 a. 2,3-dimethyl-2-butene   b. 3,5-dimethyloctane   c. 1,3,5-triethylbenzene

19.90 a. 3-ethyl-4-methyl-1-pentene   b. 2,4,6-trimethylnonane   c. 2-phenylpropane

19.91 a. CH₃C(CH₃)(CH₃)CH₂CH(CH₂CH₃)CH₂CH₂CH₂CH₃

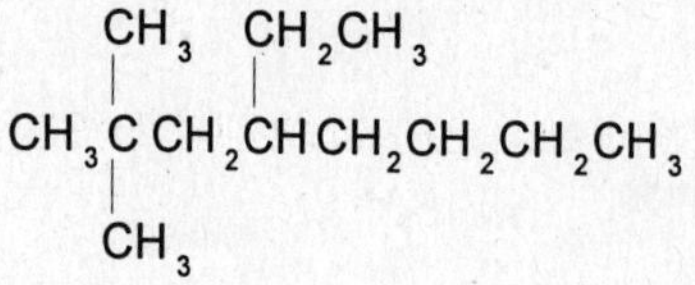

b. CH₃C(CH₃)=C(CH₃)CH₂CH₂CH₂CH₃

c. benzene ring with CH₂CH₃, CH₂CH₃, CH₂CH₃ substituents

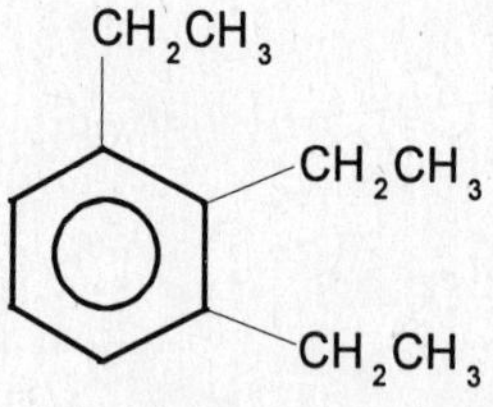

19.92 a. CH₃CH₂CH—C(CH₂CH₃)(CH₃)CH₂CH₂CH₃ with CH₃ branch

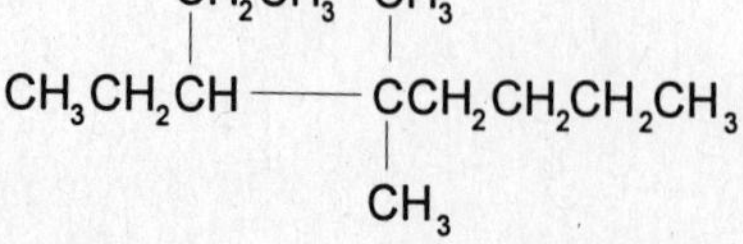

b. CH₂=C(CH₃)CH(CH₃)CH₂CH₂CH₂CH₃

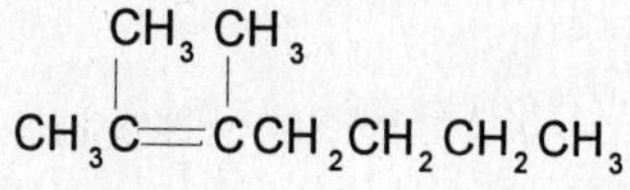

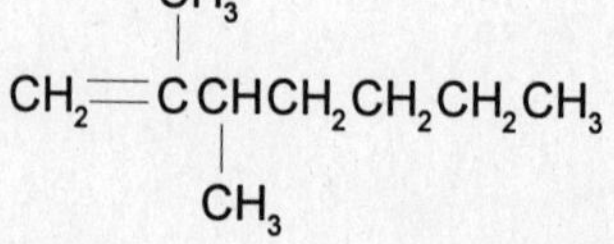

c. 1,3,5-triethylbenzene (structure shown with benzene ring bearing three CH₂CH₃ groups at positions 1, 3, 5)

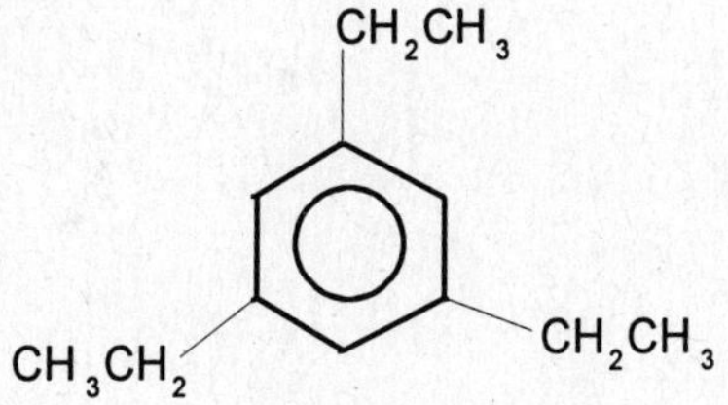

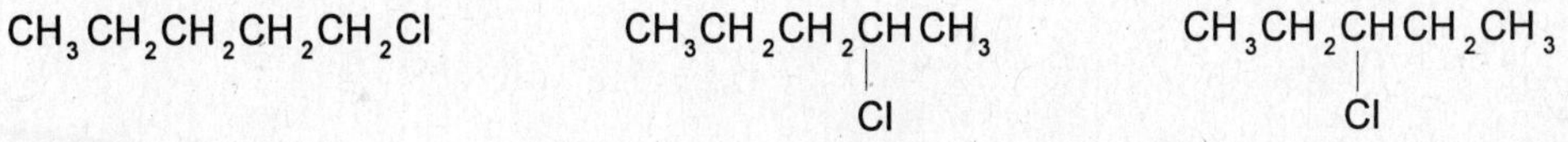

19.93 $CH_3CH_2CH_2CH_2CH_2Cl$ $CH_3CH_2CH_2CHCH_3$ $CH_3CH_2CHCH_2CH_3$
                                                  | |
                                                  Cl   Cl

19.94 $CH_3CHCH_2CH_2Cl$ $CH_3CHCHCH_3$
                | | |
                CH_3 CH_3 Cl

19.95 $CH_3CH=CHCH_3 + Br_2 \longrightarrow CH_3CHCHCH_3$
                                                          | |
                                                          Br Br

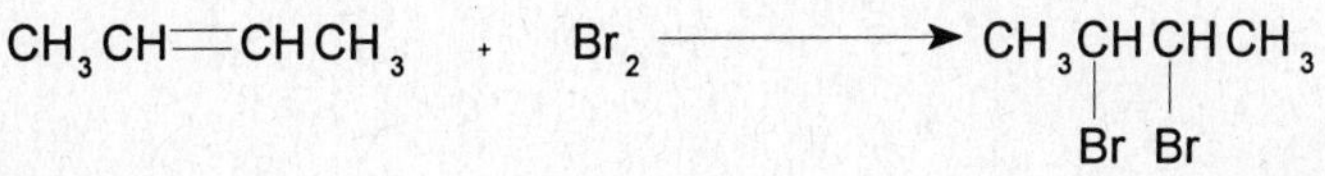

19.96 $CH_3CH_2CH=CHCH_3 + HBr \longrightarrow CH_3CH_2CH_2CHCH_3$
                                                                       |
                                                                       Br

19.97 $3\ CH_3CH_2\underset{\underset{CH_3}{|}}{\overset{\overset{CH_3}{|}}{C}}CH_2OH + 2\ Cr_2O_7^{2-} + 16\ H^+ \longrightarrow$

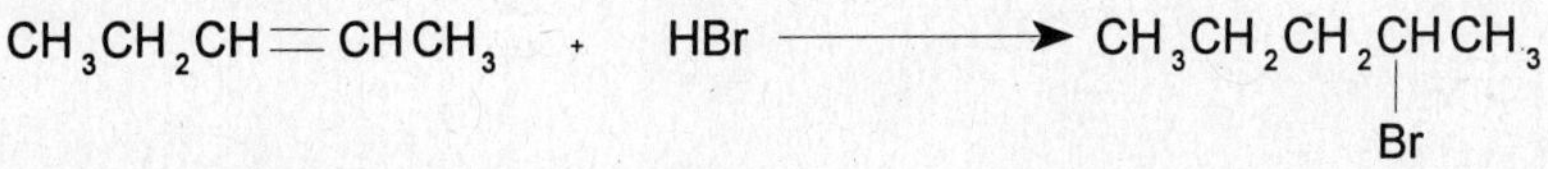

$3\ CH_3CH_2\underset{\underset{CH_3}{|}}{\overset{\overset{CH_3}{|}}{C}}\overset{O}{\overset{\|}{C}}-OH + 4\ Cr^{3+} + 11\ H_2O$

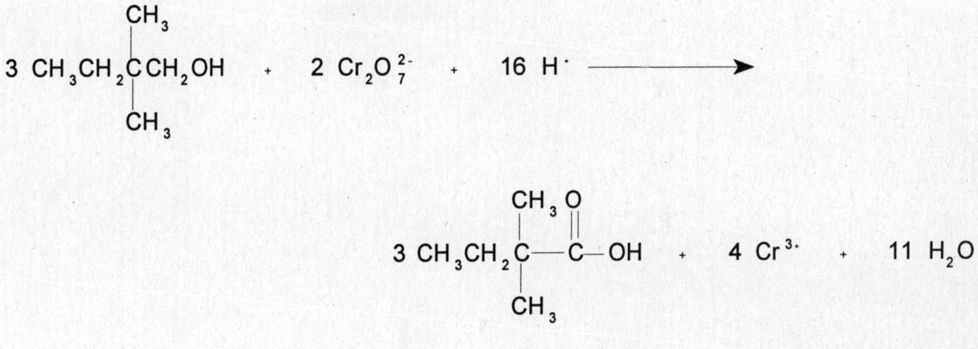

19.98 $\quad$ 3 CH$_3$CH—CH—CHCH$_3$ + Cr$_2$O$_7^{2-}$ + 8 H$^+$ $\longrightarrow$
$\qquad\qquad\quad$ | $\quad$ | $\quad$ |
$\qquad\qquad\;$ CH$_3$ OH CH$_3$

$\qquad\qquad\qquad\qquad\qquad\qquad\qquad$ CH$_3$ O CH$_3$
$\qquad\qquad\qquad\qquad\qquad\qquad\qquad\;\;$ | $\quad$ || $\;$ |
$\qquad\qquad\qquad\qquad\qquad\qquad$ 3 CH$_3$CH—C—CHCH$_3$ + 2 Cr^{3+} + 7 H$_2$O

19.99a. $\quad$ ^{4}CH$_3\,^3$C ^{2}CH═^{1}CH$_2$ $\qquad$ 3,3-dimethyl-1-butene
$\qquad\qquad\quad$ CH$_3$ (above and below on C3)

b. $\quad$ ^{5}CH$_3\,^4$CH ^{3}CH$_2\,^2$C ^{1}CH$_3$ $\qquad$ 4-methyl-2-pentanone
$\qquad\qquad$ |$\qquad\qquad$||
$\qquad\;\;$ CH$_3\qquad\quad$ O

19.100a $\quad$ ^{6}CH$_3\,^5$CH ^{4}CH ^{3}CH$_2\,^2$C ^{1}CH$_3$. $\quad$ 4,5-dimethyl-2-hexanone
$\qquad\qquad\qquad$ | $\quad$ | $\qquad\quad$ ||
$\qquad\qquad\;\;$ CH$_3$ CH$_3\qquad$ O

b. $\quad$ CH$_3$—^{3}CH ^{2}CH$_2\,^1$CH$_3$ $\qquad$ 3-methylhexane
$\qquad\qquad\quad\;$ |
$\qquad\qquad\;$ ^{4}CH$_2\,^5$CH$_2\,^6$CH$_3$

19.101 HO–C(=O)–(CH$_2$)$_2$–C(=O)–OH + H–N(H)–(CH$_2$)$_4$–N(H)–H ⟶

—C(=O)–(CH$_2$)$_2$–C(=O)–N(H)–(CH$_2$)$_4$–N(H)–C(=O)–(CH$_2$)$_2$–C(=O)–N(H)–(CH$_2$)$_4$–N(H)—

19.102 HO–C(=O)–(CH$_2$)$_2$–C(=O)–OH + HO–(CH$_2$)$_2$–OH ⟶

—C(=O)–(CH$_2$)$_2$–C(=O)–O–(CH$_2$)$_2$–O–C(=O)–(CH$_2$)$_2$–C(=O)–O–(CH$_2$)$_2$–O—

20. BIOCHEMISTRY

■ Solutions to Exercises

20.1 Glycine and phenylalanine can react to form two possible structures. The structural formulas are

glycine + phenylalanine

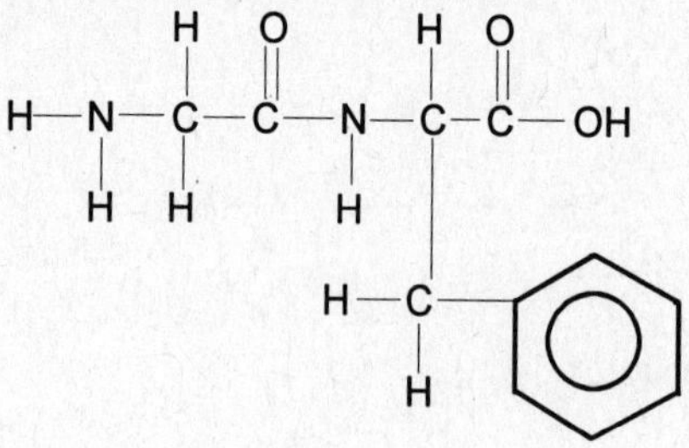

phenylalanine + glycine

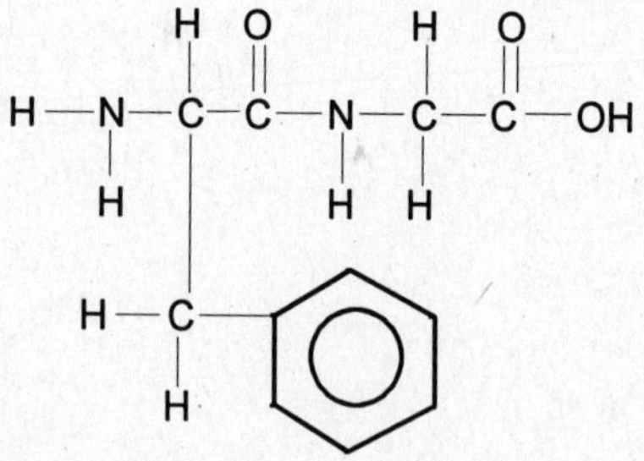

20.2

<u>aldose</u>

```
      O
      ‖
      C — H
      |
  H — C — OH
      |
  H — C — OH
      |
  H — C — OH
      |
  H — C — OH
      |
  H — C — OH
      |
  H — C — OH
      |
      H
```

<u>ketose</u>

```
      H
      |
  H — C — OH
      |
      C = O
      |
  H — C — OH
      |
  H — C — OH
      |
  H — C — OH
      |
  H — C — OH
      |
      H
```

20.3 The segment of DNA that has the base sequence T-A-C-G will have the following base sequence in the complementary chain: A-T-G-C.

20.4 The amino acid sequence Glu-Met-Tyr can have one of four different sequences in the messenger chain. These are

   G-A-A-A-U-G-U-A-C   G-A-A-A-U-G-U-A-U
   G-A-G-A-U-G-U-A-C   G-A-G-A-U-G-U-A-U

■ Answers to Questions to Test Your Reading

20.1 The two types of cells are <u>procaryote</u> and <u>eucaryotes</u>. Procaryote are single-celled organisms, such as bacteria. Eucaryotes are all multi-cell organisms, such as plants and animals, as well as some single-celled organisms, such as yeast. The main difference between the two is the presence in eucaryotes of <u>organelles</u>, which are parts of the cell having distinct functions, such as the nucleus.

20.2 The <u>nucleus</u> directs the workings of the cell. It contains the information for the preparation of proteins and also the information that is passed to the next generation. <u>Mitochondria</u> are organelles that produce the energy of the cell from food molecules, such as glucose.

20.3 The four classes of biological molecules are proteins, nucleic acids, carbohydrates, and lipids. Proteins are chain polymers composed of amino acids, which are organic molecules containing an amine group and a carboxylic acid group. Carbohydrates are either simple sugars or else compounds consisting of two or more simple sugar units. They have a carbonyl group and two or more hydroxyl groups. Nucleic acids are polymers composed of nucleotide units. These are either deoxyribose or ribose to which a phosphate unit and a nitrogen-containing base are attached. Lipids are biological substances that are soluble in organic solvents. These include steroids and fats.

20.4 Proteins can function as catalysts and are called _enzymes_. They also function as molecular transporters. An example is _lipoproteins_. They also function as structural support as in skin and bone. Or, they function in movement as in muscle tissue.

20.5 Examples of carbohydrates include the sugars, such as glucose (or dextrose) and sucrose. Starch is an example of a glucose polymer, as in cellulose.

20.6 The two types of nucleic acids are deoxyribonucleic acid (DNA) and ribonucleic acid (RNA). DNA is the repository of a cell's genetic information. The RNA is a messenger molecule that forms a template for the synthesis of a given protein molecule.

20.7 The lipids include the following two groups. The _steroids_ are fused-ring compounds that include sex hormones and cholesterol. The fats, or _triacylglycerols_ are molecules that consist of glycerols bonded through ester linkages to three fatty acids. Another type of fat is a _phospholipid_, which is a constituent of cell membranes.

20.8 $H_2N-\underset{H}{\overset{CH_2OH}{C}}-COOH$ the amino acid is _serine_

20.9 $CH_3-\underset{NH_2}{\overset{H}{C}}-\underset{H}{\overset{H}{C}}-COOH$ a β-amino acid

20.10 Ala-Gly

$\underset{H}{\overset{H}{N}}-\underset{H}{\overset{CH_3}{C}}-\overset{O}{C}-\underset{H}{N}-\underset{H}{\overset{H}{C}}-\overset{O}{C}-OH$

It differs from Gly-Ala in the location of the CH_3 group.

20.11 Tripeptides of Phe, Ala, and Gly are   Phe-Ala-Gly   Ala-Phe-Gly   Gly-Ala-Phe
Phe-Gly-Ala   Ala-Gly-Phe   Gly-Phe-Ala

20.12 The <u>primary structure</u> of a protein or a peptide refers to the order, or sequence, of the amino acid units in the protein or peptide molecule, and also the locations of any disulfide bonds that might be present.

20.13

$$\begin{array}{c}\text{H}\quad\text{H}\quad\text{O}\\ \text{H}-\text{N}-\text{C}-\text{C}-\text{OH}\\ |\\ \text{CH}_2\\ |\\ \text{SH}\end{array}\quad+\quad\begin{array}{c}\text{SH}\\ |\\ \text{CH}_2\\ |\\ \text{H}-\text{N}-\text{C}-\text{C}-\text{OH}\\ \text{H}\quad\text{H}\quad\text{O}\end{array}\quad+\ (O)\ \longrightarrow\quad\begin{array}{c}\text{H}\quad\text{H}\quad\text{O}\\ \text{H}-\text{N}-\text{C}-\text{C}-\text{OH}\\ |\\ \text{CH}_2\\ |\\ \text{S}\\ |\\ \text{S}\\ |\\ \text{CH}_2\\ |\\ \text{H}-\text{N}-\text{C}-\text{C}-\text{OH}\\ \text{H}\quad\text{H}\quad\text{O}\end{array}\quad+\ \text{H}_2\text{O}$$

20.14 Digestion is the reverse process of building proteins from amino acids. Enzymes present in digestive fluids catalyze the breaking of the peptide bonds to yield amino acids.

20.15 Although kidney beans are incomplete proteins in that they do not have all the essential amino acids, when combined with other protein foods, such as rice, the combination of foods has all the essential amino acids and is a complete protein diet.

20.16 In globular proteins, the coils or helixes of the proteins fold themselves into compact spherical shapes (or globules) in which the nonpolar side chains of the protein, which are not attracted to water, are inside the globule. The polar groups are on the outside of the globule, where they are attracted to water. The forces responsible for the action are the dispersion forces between the nonpolar portions of the chain on the inside of the globule and also dipole-dipole forces between the polar groups and the water on the outside of the globule.

20.17 When a protein is <u>denatured</u>, it loses its three-dimensional shape through the unfolding and uncoiling of the protein as a result of the breaking of the weak forces, such as hydrogen bonding, that holds the protein in its normal three-dimensional shape. The normal biochemical function is lost. However, if the denatured protein is purified and oxidized, it reverses its shape and activity.

20.18 The term carbohydrate arises from the general molecular formulas of these substances, $C_m(H_2O)_n$. The actual structures of these compounds is very different from this. They are really polyhydroxy ketones or aldehydes, or else a substance that yields such compounds if the carbohydrate hydrolyzes with water.

20.19

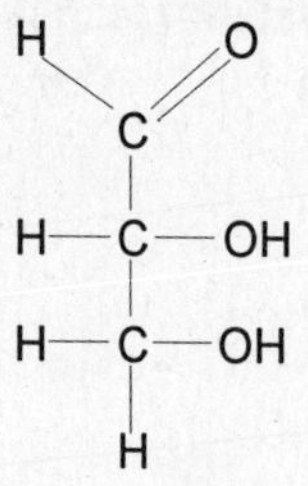

glyceraldehyde, an aldose 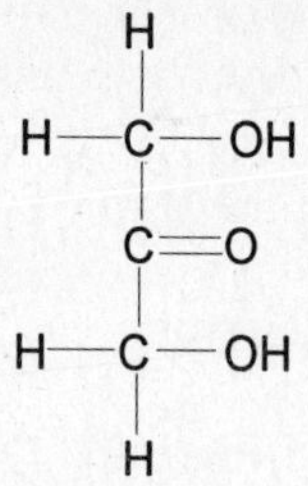   dihydroxyacetone, a ketose

20.20 A monosaccharide is a simple sugar such as glucose, fructose, ribose, and 2-deoxyribose. An oligosaccharide is a short polymer of two to ten monosaccharide units. An example is sucrose, which contains a glucose and fructose units. A polysaccharide is a polymer containing many monosaccharide units. Starch and cellulose are examples of glucose polymers.

20.21 The final compound obtained in the complete digestion of starch is the monosaccharide glucose. This is a hydrolysis reaction, or a reaction with water.

20.22 Nucleic acids are the molecular carriers of genetic information. The sugar constituent of DNA, the primary carrier, is 2-deoxyribose.

20.23 The four bases contained in DNA are cytosine, thymine, adenine, and guanine. In RNA, the bases are the same except uracil replaces thymine.

20.24 ATP is used by organisms as an immediate source of energy. When glucose is broken down in a cell, the energy obtained is stored in the ATP molecule by the following reaction

ADP + phosphate ion + energy ⟶ ATP + H_2O

During biochemical processes that require energy, the reverse process occurs.

20.25 DNA consists of a strand of two complementary polynucleotide chains twisted together into a double helix. The sequence of different bases on the chain is a code representing the sequence of amino acids in a series of protein molecules. During cell division, the coil unravels and duplicates, producing two cells, each with its own DNA double helix.

20.26 The double helix structure is due to complementary base pairing, which is pairing through hydrogen bonding, of certain bases in DNA with others. Certain base pairs, such as adenine and thymine, form multiple hydrogen bonds with each other. One of the polynucleotide strands in a length of DNA is the complement of the other strand.

20.27 RNA is much smaller than DNA. This is because the molecules, such as <u>messenger RNA</u> (mRNA) are complementary copies of a segment of DNA for a particular protein. DNA contains the sequences for many proteins.

20.28 The flow of protein information begins with <u>transcription</u>, in which the information for the sequence of amino acids in a protein is transferred from a DNA chain in the cell nucleus to a messenger RNA molecule. The messenger RNA molecule migrates from the cell nucleus to the cytoplasm, where it serves as a template on which the amino acids are assembled in a process called <u>translation</u>.

20.29 Phenylalanine (Phe) has two possible codons: UUU and UUC.

20.30 The <u>transfer RNA</u> (tRNA) molecules are smaller than mRNA molecules and actually perform the translation of the RNA base code. Each codon, or base triplet, on the mRNA chain links only to a tRNA molecule having the complementary anticodon. Each tRNA molecule in turn carries a particular amino acid.

20.31 A <u>lipid</u> is a biological substance belonging to one of several structurally different classes of substances that are soluble in organic solvents, such as chloroform, $CHCl_3$. Two examples are <u>triacylglycerols</u> (fats) and <u>phospholipids</u>.

20.32 When a fat (triacylglycerol) is heated with a solution of sodium hydroxide (lye), it is hydrolyzed, yielding glycerol and the sodium salts of fatty acids, or soap. The process is called <u>saponification</u>.

20.33 A soap is a long molecule with a nonpolar end and a carboxylate ion end, which is polar. An example is the stearate ion, $CH_3(CH_2)_{16}CO_2^-$. When these ions are in water, they form <u>micelles</u> that absorb the oil and grease into the hydrocarbon (nonpolar) centers while the polar ends are outward toward the water, and can be washed away.

20.34 Phospholipids form bilayers, or sheets two molecules thick, because the two fatty acid groups are bulky and prevent the formation of more complicated micelles.

■ Solutions to Practice Problems

20.35   <u>alanine</u>   <u>β-amino acid</u>

$$CH_3\underset{\underset{NH_2}{|}}{C}HCOOH \qquad \qquad \underset{\underset{NH_2}{|}}{C}H_2CH_2COOH$$

20.36 valine   β-amino acid

$(CH_3)_2CHCHCOOH$ $(CH_3)_2CCH_2COOH$
        | |
       NH_2 NH_2

20.37

$H-N-\underset{\underset{CH_3}{|}}{\overset{\overset{H}{|}}{C}}-\overset{\overset{O}{\|}}{C}-OH$ + $H-N-\underset{\underset{\underset{\underset{}{}}{CH_2-C_6H_5}}{|}}{\overset{\overset{H}{|}}{C}}-\overset{\overset{O}{\|}}{C}-OH$ ⟶

$H-N-\underset{\underset{CH_3}{|}}{\overset{\overset{H}{|}}{C}}-\overset{\overset{O}{\|}}{C}-N-\underset{\underset{CH_2-C_6H_5}{|}}{\overset{\overset{H}{|}}{C}}-\overset{\overset{O}{\|}}{C}-OH$ + H_2O

$H-N-\underset{\underset{CH_2-C_6H_5}{|}}{\overset{\overset{H}{|}}{C}}-\overset{\overset{O}{\|}}{C}-OH$ + $H-N-\underset{\underset{CH_3}{|}}{\overset{\overset{H}{|}}{C}}-\overset{\overset{O}{\|}}{C}-OH$ ⟶

$H-N-\underset{\underset{CH_2-C_6H_5}{|}}{\overset{\overset{H}{|}}{C}}-\overset{\overset{O}{\|}}{C}-N-\underset{\underset{CH_3}{|}}{\overset{\overset{H}{|}}{C}}-\overset{\overset{O}{\|}}{C}-OH$ + H_2O

The two product molecules have completely different structures.

20.38

$H_2N-CH(CH(CH_3)_2)-COOH$ + $H_2N-CH(CH_2CH_2SCH_3)-COOH$ ⟶

$H_2N-CH(CH(CH_3)_2)-CO-NH-CH(CH_2CH_2SCH_3)-COOH$ + H_2O

$H_2N-CH(CH_2CH_2SCH_3)-COOH$ + $H_2N-CH(CH(CH_3)_2)-COOH$ ⟶

$H_2N-CH(CH_2CH_2SCH_3)-CO-NH-CH(CH(CH_3)_2)-COOH$ + H_2O

The two product molecules have completely different structures.

20.39 Glycine (Gly), alanine (Ala), methionine (Met), and leucine (Leu)

Gly-Ala-Met-Leu	Ala-Gly-Met-Leu	Met-Gly-Ala-Leu	Leu-Met-Gly-Ala
Gly-Ala-Leu-Met	Ala-Gly-Leu-Met	Met-Gly-Leu-Ala	Leu-Met-Ala-Gly
Gly-Met-Ala-Leu	Ala-Leu-Gly-Met	Met-Ala-Gly-Leu	Leu-Gly-Met-Ala
Gly-Met-Leu-Ala	Ala-Leu-Met-Gly	Met-Ala-Leu-Gly	Leu-Gly-Ala-Met
Gly-Leu-Ala-Met	Ala-Met-Leu-Gly	Met-Leu-Ala-Gly	Leu-Ala-Gly-Met
Gly-Leu-Met-Ala	Ala-Met-Gly-Leu	Met-Leu-Gly-Ala	Leu-Ala-Met-Gly

The amine groups are on the left and the carboxylic acid groups are on the right.

20.40 Tyrosine (Tyr), glycine (Gly), phenylalanine (Phe), and leucine (Leu)

Tyr-Gly-Phe-Leu	Gly-Tyr-Phe-Leu	Phe-Tyr-Gly-Leu	Leu-Phe-Tyr-Gly
Tyr-Gly-Leu-Phe	Gly-Tyr-Leu-Phe	Phe-Tyr-Leu-Gly	Leu-Phe-Gly-Tyr
Tyr-Phe-Gly-Leu	Gly-Leu-Tyr-Phe	Phe-Gly-Tyr-Leu	Leu-Tyr-Phe-Gly
Tyr-Phe-Leu-Gly	Gly-Leu-Phe-Tyr	Phe-Gly-Leu-Tyr	Leu-Tyr-Gly-Phe
Tyr-Leu-Gly-Phe	Gly-Phe-Leu-Tyr	Phe-Leu-Gly-Tyr	Leu-Gly-Tyr-Phe
Tyr-Leu-Phe-Gly	Gly-Phe-Tyr-Leu	Phe-Leu-Tyr-Gly	Leu-Gly-Phe-Tyr

The amine groups are on the left and the carboxylic acid groups are on the right.

20.41 aldoses:

5-carbon

$$\begin{array}{c} O \\ \parallel \\ C-H \\ | \\ H-C-OH \\ | \\ H-C-OH \\ | \\ H-C-OH \\ | \\ H-C-OH \\ | \\ H \end{array}$$

6-carbon

$$\begin{array}{c} O \\ \parallel \\ C-H \\ | \\ H-C-OH \\ | \\ H-C-OH \\ | \\ H-C-OH \\ | \\ H-C-OH \\ | \\ H-C-OH \\ | \\ H \end{array}$$

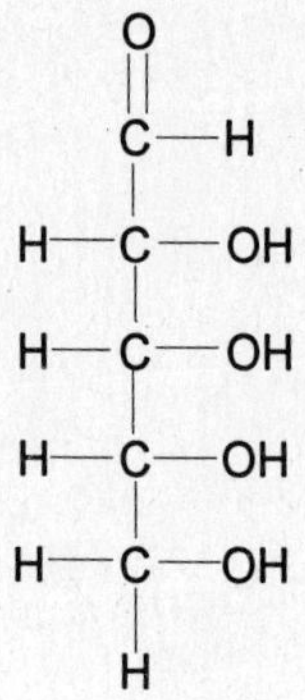

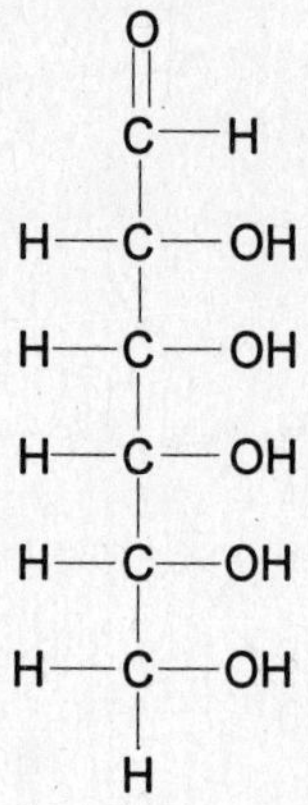

20.42 ketoses:

5-carbon

$$\begin{array}{c} H \\ | \\ H-C-OH \\ | \\ C=O \\ | \\ H-C-OH \\ | \\ H-C-OH \\ | \\ H-C-OH \\ | \\ H \end{array}$$

6-carbon

$$\begin{array}{c} H \\ | \\ H-C-OH \\ | \\ C=O \\ | \\ H-C-OH \\ | \\ H-C-OH \\ | \\ H-C-OH \\ | \\ H-C-OH \\ | \\ H \end{array}$$

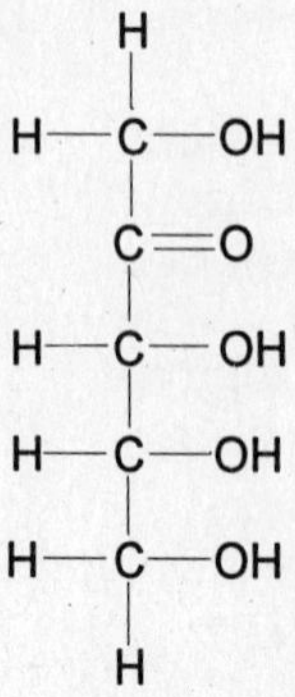

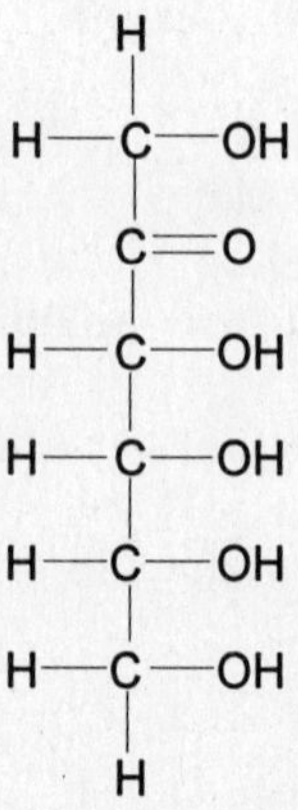

20.43 $C_{12}H_{22}O_{11}$ + H_2O $\xrightarrow{H^+}$ $C_6H_{12}O_6$ + $C_6H_{12}O_6$
Lactose   Glucose   Galactose

20.44 $C_{12}H_{22}O_{11}$ + H_2O $\xrightarrow{H^+}$ $C_6H_{12}O_6$ + $C_6H_{12}O_6$
Maltose   Glucose   Glucose

20.45 A-T-G. The base sequence of the complementary chain is T-A-C.

20.46 G-A-T-A. The base sequence of the complementary chain is C-T-A-T.

20.47 Ala-Gly-Phe. There are many possible base sequences. One is: GCAGGAUUU. (See Table 20-3 for the various codons)

20.48 Leu-Met-Gly. There are many possible base sequences. One is: CUAAUGGGA. (See Table 20-3 for the various codons)

20.49 For a GCAGGAUUU sequence in m-RNA, the corresponding base sequence in DNA would be CGTCCTAAA.

20.50 For a CUAAUGGGA sequence in m-RNA, the corresponding base sequence in DNA would be GATTACCCT.

20.51

```
     H   O
     |   ||
H—C—O—C—(CH₂)₁₀CH₃
     |
     |   O
     |   ||
H—C—O—C—(CH₂)₁₆CH₃
     |
     |   O
     |   ||
H—C—O—C—(CH₂)₇CH=CH(CH₂)₇CH₃
     |
     H
```

20.52

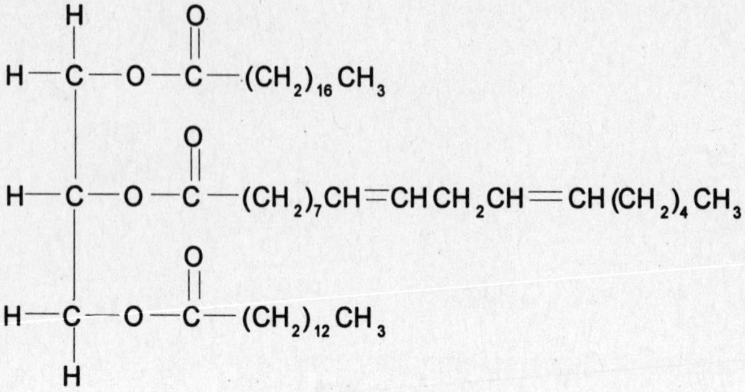

$$\begin{array}{l} H-\overset{H}{\underset{}{C}}-O-\overset{O}{\underset{}{C}}-(CH_2)_{16}CH_3 \\ H-\overset{}{\underset{}{C}}-O-\overset{O}{\underset{}{C}}-(CH_2)_7CH=CHCH_2CH=CH(CH_2)_4CH_3 \\ H-\overset{}{\underset{H}{C}}-O-\overset{O}{\underset{}{C}}-(CH_2)_{12}CH_3 \end{array}$$

20.53

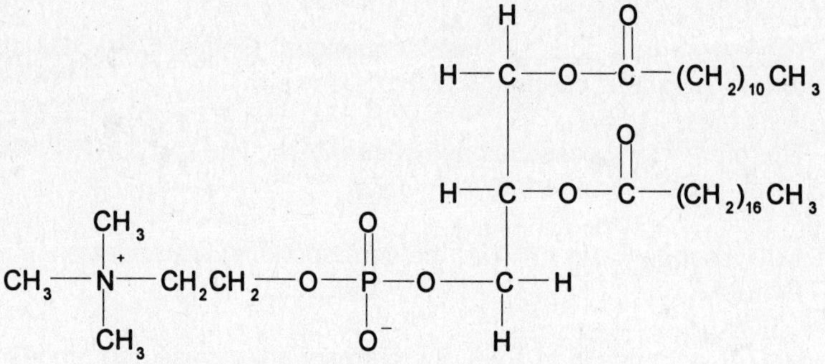

$$CH_3-\overset{CH_3}{\underset{CH_3}{\overset{+}{N}}}-CH_2CH_2-O-\overset{O}{\underset{O^-}{P}}-O-\overset{H}{\underset{H}{C}}-H$$

with the glyceride portion:
$$\begin{array}{l} H-C-O-C-(CH_2)_{10}CH_3 \\ H-C-O-C-(CH_2)_{16}CH_3 \end{array}$$

20.54

$$CH_3-\overset{CH_3}{\underset{CH_3}{\overset{+}{N}}}-CH_2CH_2-O-\overset{O}{\underset{O^-}{P}}-O-\overset{H}{\underset{H}{C}}-H$$

with the glyceride portion:
$$\begin{array}{l} H-C-O-C-(CH_2)_{16}CH_3 \\ H-C-O-C-(CH_2)_7CH=CHCH_2CH=CH(CH_2)_4CH_3 \end{array}$$

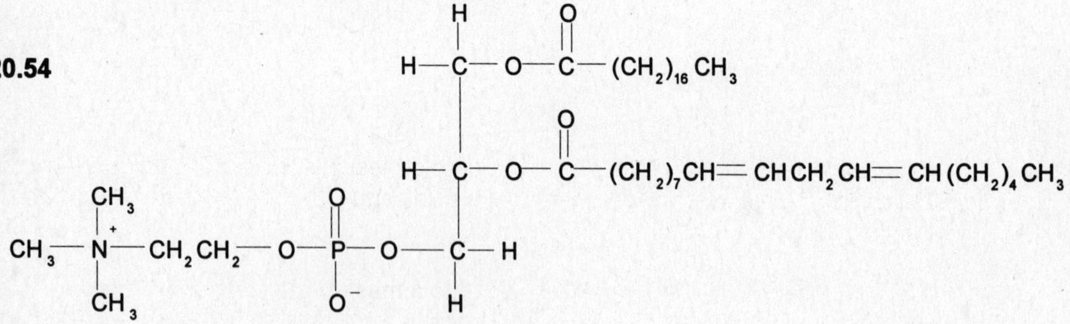

■ Solutions to Additional Problems

20.55 alanine (Ala), glycine (Gly), and valine (Val). There are six possibilities.

Ala-Gly-Val	Gly-Ala-Val	Val-Gly-Ala
Ala-Val-Gly	Gly-Val-Ala	Val-Ala-Gly

20.56 leucine (Leu), glycine (Gly), and methionine (Met). There are six possibilities.

| Leu-Gly-Met | Gly-Met-Leu | Met-Gly-Leu |
| Leu-Met-Gly | Gly-Leu-Met | Met-Leu-Gly |

20.57 The disulfide bond forms between the cystine units.

```
Glu — Gln — Cys — Thr
              |      |
              S     His
              |      |
              S     Ser
              |      |
His — Leu — Cys — Gly
```

20.58 The disulfide bond forms between the cystine units.

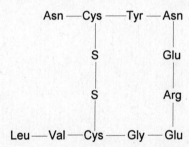

20.59 DNA base sequence TTGCTTCAA would copy to a messenger RNA as AACGAAGUU. This base sequence corresponds to the amino acid sequence Asn-Glu-Val. (See Table 20-3 for the various codons)

20.60 DNA base sequence CTATGCTAA would copy to a messenger RNA as GAUACGAUU. This base sequence corresponds to the amino acid sequence Asp-Thr-Ile. (See Table 20-3 for the various codons)

APPENDIX A. MATHEMATICAL SKILLS

■ Solutions to Exercises

1.
   a. $5.2 \times 10^9 + 6.3 \times 10^9 = 11.5 \times 10^9 = 1.15 \times 10^{10}$

   b. $3.142 \times 10^6 + 2.8 \times 10^4 = 3.142 \times 10^6 + 0.028 \times 10^6 = 3.170 \times 10^6$

   c. $3.142 \times 10^6 - 2.8 \times 10^4 = 3.142 \times 10^6 - 0.028 \times 10^6 = 3.114 \times 10^6$

   d. $4.1 \times 10^8 + 4 \times 10^7 = 4.1 \times 10^8 + 0.4 \times 10^8 = 4.5 \times 10^8$

   e. $4.1 \times 10^8 - 4 \times 10^7 = 4.1 \times 10^8 - 0.4 \times 10^8 = 3.7 \times 10^8$

2.
   a. $5.4 \times 10^{-7} \times 1.8 \times 10^8 = (5.4 \times 1.8) \times 10^1 = 9.\underline{7}2 \times 10^1 = 9.7 \times 10^1$

   b. $\dfrac{3.0 \times 10^{-6}}{6.0 \times 10^3} = \dfrac{3.0}{6.0} \times 10^{-9} = 0.50 \times 10^{-9} = 5.0 \times 10^{-10}$

   c. $\dfrac{3.0 \times 10^6}{6.0 \times 10^{-3}} = \dfrac{3.0}{6.0} \times 10^9 = 0.50 \times 10^9 = 5.0 \times 10^8$

   d. $6.1 \times 10^9 \times 2.3 \times 10^2 = (6.1 \times 2.3) \times 10^{11} = 1\underline{4}.03 \times 10^{11} = 1.4 \times 10^{12}$

   e. $\dfrac{2.5 \times 10^4}{5.0 \times 10^6} = \dfrac{2.5}{5.0} \times 10^{-2} = 0.50 \times 10^{-2} = 5.0 \times 10^{-3}$

3.
   a. $b = \dfrac{V}{T}$
   b. $T = \dfrac{PV}{nR}$
   c. $T_2 = \dfrac{V_2 T_1}{V_1}$

   d. $\text{volume} = \dfrac{\text{mass}}{\text{density}}$
   e. $°F = \dfrac{9}{5} °C + 32$

4. At point C, the volume of the gas is 1 L, and the pressure is 4 atm.

APPENDIX B. USING YOUR CALCULATOR

■ Solutions to Exercises

1.   a. $5.67 \times 10^{-3} + 2.1 \times 10^{-4} = 5.88 \times 10^{-3}$

       b. $3.0 \times 10^{7} - 3 \times 10^{6} = 2.7 \times 10^{7}$

       c. $4.3 \times 10^{-2} \times 9 \times 10^{10} = \underline{3}.87 \times 10^{9} = 4 \times 10^{9}$

       d. $\dfrac{7.2 \times 10^{-5}}{3.6 \times 10^{-7}} = 2.0 \times 10^{2}$

       e. $\dfrac{4.4 \times 10^{3} \times 1.7 \times 10^{2}}{5.1 \times 10^{-3} \times 4.4 \times 10^{-5}} = 3.3 \times 10^{8}$

       f. $\dfrac{2.5 \times 10^{-2} + 1.0 \times 10^{-1}}{5.5 \times 10^{-2} - 5.0 \times 10^{-3}} = \dfrac{1.25 \times 10^{-1}}{5.0 \times 10^{-2}} = 2.5$

       g. $1.1 \times 10^{2} + 2.0 \times 10^{3} \times 1.5 \times 10^{-2} - \dfrac{3.6 \times 10^{9}}{2.4 \times 10^{7}} = -1.0 \times 10^{1}$